Statistics and Econometrics

A Problem-Solving Text

Statistics and Econometrics

A Problem-Solving Text

Barry R. Chiswick

Queens College
City University of New York
and
Stephen J. Chiswick
Bronx Community College
City University of New York

University Park Press

Baltimore · London · Tokyo

UNIVERSITY PARK PRESS

International Publishers in Science and Medicine
Chamber of Commerce Building
Baltimore, Maryland 21202

Copyright © 1975 by University Park Press

Printed in the United States of America

All rights, including that of translation into other languages,
reserved. Photomechanical reproduction (photocopy, microcopy) of
this book or parts thereof without special permission of the pub-
lisher is prohibited.

Type composed at Mono of Maryland, Baltimore, Maryland

Library of Congress Cataloging in Publication Data

Chiswick, Barry R
  Statistics and econometrics.

  Includes index.

  1. Statistics. 2. Econometrics. 3. Regression
analysis. I. Chiswick, Stephen J., joint author.
II. Title.
HA29.C5544   519.5   75-1257
ISBN 0-8391-0694-7

To our mother Yetta,
and the memory
of our father Abraham.

Contents

Preface

Recent developments in data collection and processing have altered considerably the nature of data analysis and decision making in business and the social sciences. The empirical testing of hypotheses has become a major aspect of decision making and research.

The typical method for teaching applied statistics is to combine lectures (often to large classes) with the reading of textbooks. Unfortunately, most textbooks devote little attention to the techniques of empirical analysis. Although the student may learn statistical theory when it comes to actually using empirical analysis, he or she is all too often inexperienced. What students need is practice solving problems and a text that shows, step by step, how problems can be solved and why they are solved in a particular manner.

The introduction to each chapter in this book contains discussions of aspects of statistical theory. These discussions serve to clarify the theoretical points and empirical techniques, thus providing the reader with a deeper understanding of the relevance of statistical theory. This is followed by a set of problems that can be solved by the statistical techniques typically taught in a first-year course in statistics and econometrics. Annotated solutions to these problems are presented so that the student can learn the *how* and *why* of solving.

A knowledge of advanced mathematics is generally not needed in order to learn the techniques of statistical analysis. Far too often, students lose sight of the important basic principles when the analysis is combined with mathematical tools with which they have only a superficial acquaintance. Thus this book does *not* use matrix algebra or calculus. It does, however, assume a familiarity with elementary (ninth-grade) algebra.

Part A develops the basic tools of statistical analysis. Assuming that the reader has no prior knowledge of statistics, we start with descriptive statistics in Chapter 1. Chapter 2 is an elementary treatment of probability theory. Chapters 3 and 4 develop the principles of hypothesis testing and apply them to analyses of means, proportions, variances, and frequency distributions. Part B is devoted to one of the most important techniques for analyzing data—regression analysis. Chapters 5 and 6 are explorations of the single-equation simple and multiple regression models, respectively. Chapter 7 develops procedures for estimating simultaneous systems of equations.

This book can be used in either a one- or a two-semester course in statistics and/or econometrics, either as a textbook to supplement lectures or in conjunction with more traditional texts.

We are indebted to the Literary Executor of the late Sir Ronald A. Fisher, F.R.S., to Dr. Frank Yates, F.R.S., and to Longman Group Ltd., London, for permission to reprint Tables 5, 11, and 12 from their book *Statistical Tables for Biological, Agricultural and Medical Research*.

We are also indebted to our students in statistics, and to many of our colleagues for their comments and encouragement on this book.

Introduction

Statistics and econometrics are a body of procedures for obtaining and analyzing data in order to generate knowledge on which to base decisions. The procedures are limited to those things that can be described numerically, on either a cardinal or an ordinal scale. Business and social science decision making generally cannot be based on controlled experiments, and predictions about the future must therefore be made on the basis of past (historical) data.

The field of *applied statistics* consists of *descriptive statistics* and *analytical statistics*. Descriptive statistics involves the condensation of sets (or arrays) of data into a smaller number of summary measures. For example, data on last year's income of the American labor force would consist of over 100 million observations. By themselves, these data are too plentiful to be of use to us. However, if we organize the data by creating a frequency distribution or by calculating the average level or inequality of income in the American labor force, we do obtain useful information. Chapter 1 is devoted to descriptive statistics.

Analytical statistics is concerned with *statistical populations* and *samples*. A statistical population consists of the sum of all possible observations of the same kind, e.g., members of the American labor force, or the number of years in American history. A sample consists of the members of the population that have been observed. For example, the 1970 Census of Population asked detailed questions on income from only 5 percent of the labor force, or data on the schooling of the population of the United States exist for only a few decades.

The basis of statistical analysis is *random sampling*. If each element of a population has the same chance or probability of being selected for inclusion in a sample, the sampling procedure and the resulting sample are said to be random. Random sampling and probability theory permit a computation of how likely it is that a particular sample from a population with hypothesized characteristics will be obtained.

The characteristics of a population (e.g., average income) are unknown but sample values are known. The sample values, however, vary from sample to sample. The pattern of variation of sample values depends on the size of the sample and on the manner in which the sample is selected. Samples are generally substantially smaller than the size of the population. In some cases, the testing of elements in a population may destroy the item (e.g., flashbulbs). Also, some populations have an infinite size (e.g., a study of rainfall on Manhattan Island in past Julys has an infinite population). Perhaps the most important determinant of sample size, however, is its cost. The increased information may not be worth the greater cost of a larger sample. For example, today's average daily temperature can be estimated by taking readings every hour on the hour—24 observations—or by taking readings every second—86,400 observations. For most purposes, the value of this greater accuracy is small, but the extra cost is large. It will be shown below that random samples that are a small proportion of the elements in a large population can produce very reliable results.

There are two forms of analytical statistics, *probability* and *inference*. *Probability* uses the scientific principle of deduction; on the basis of assumptions about the population a statement is made as to the probable composition of a sample. For example, if we know the proportion of families in the United States with two cars, probability theory can tell us how likely it is that we will obtain a sample in which, say, not more than 20 percent of the families have two cars. Chapter 2 is devoted to probability.

Chapters 3 and 4 are concerned with *statistical inference*, which is based on the principle of induction. Statistical inference uses knowledge of a sample to obtain information about the population. If we know what proportion of a sample of families own two cars, we can determine how likely it is that 20 percent of the families in the entire population own two cars.

Statistical inference is used to test hypotheses and to estimate the characteristics of a population.

A *hypothesis* is an assumption about a characteristic of a population. For example, one hypothesis is that the average income in the American labor force last year was $8,000. Without sampling the entire population it is impossible to *prove* that the hypothesis is true. However, statistical inference does provide a means of testing whether sample data are consistent with a hypothesis, or whether the data are such that it is "too unlikely" that the hypothesis is valid.

Sample data can also be used to estimate the characteristics of a population (e.g., average income). A *confidence interval* uses sample data to compute a range of values which "most probably" contain the true, but unknown, population value. Without sampling the entire population we can never be 100 percent certain that the population value is within the specified interval.

Chapter 3 develops the principles of hypothesis testing and confidence intervals, and applies these principles to analyses of mean values and proportions in populations. Hypothesis testing and confidence intervals for the variance in a population are presented in Chapter 4, which also includes procedures for testing whether one or more sample frequency distributions come from a hypothesized population and whether the samples are from the same population.

Econometrics is the application of mathematical and statistical techniques to analyses of economic variables. Empirical studies of economic, social, and business problems usually require multivariate analysis, that is, analysis which involves several variables. Econometrics has become strongly identified with one particular type of multivariate analysis called *regression analysis*. Regression analysis is a statistical procedure for finding the "best" fitting line or equation relating a dependent variable to one or more explanatory (independent) variables. Part B, Chapters 5 through 7, develops the computation and interpretation of the regression equations. Procedures are presented for identifying and correcting difficulties that arise when one or more of the assumptions of the regression model are violated. Regression analysis is used to test hypothesized effects of explanatory variables on the dependent variable and to make predictions about the magnitude of the dependent variable. Chapter 5 is concerned with the *simple regression model* in which there is only one explanatory variable. In Chapter 6 this is extended to the *multiple regression model* of more than one explanatory variable. Under certain circumstances it is inappropriate to use a single equation, and it is necessary to estimate a set of *simultaneous equations*; this is dealt with in Chapter 7.

Part A

Descriptive Statistics, Probability, and Hypothesis Testing

This part presents the basic building blocks for the statistical analysis of data and for statistical decision making. The principles and techniques demonstrated here are widely used for these purposes. In addition, they serve as essential inputs into more advanced statistical procedures, such as regression analysis, which is presented in Part B.

Chapter 1

Descriptive Statistics

Suppose we are interested in the distribution of income of families in the United States, and we collect a sample of 1,000 observations. By themselves, the 1,000 separate observations of family income are clearly too much information to be of use. However, if we can convert the data into a few summary measures that describe particular characteristics of the distribution of income among the families, we can obtain useful, manageable information. *Descriptive statistics* involves the condensation of sets of data into a smaller number of summary measures. Descriptive statistics includes frequency distributions, measures of central tendency, measures of dispersion, and measures of the shape of a distribution.

A *frequency distribution* can be used to bring some order to an array of observations. For each of the values of the variable under study, a frequency distribution indicates the number of times (frequency) that value of the variable appears in the data. For example, suppose a sample of 1,000 families indicates that the lowest income is $5,550 and the highest income is $15,300. Using intervals of $1,000, we can express the observations as a frequency distribution with eleven intervals (e.g., first interval: $5,500 to $6,500, second interval: $6,500 to $7,500, etc.), with an indication of the number of families with incomes in each interval. At a glance we can learn how many families have incomes less than $9,000, or more than $12,000, and so on.

Measures of central tendency provide information about the location of the center of a distribution. The *arithmetic mean*—the sum of the values of the observations divided by the number of observations—is the most common measure of central tendency. The *modal value* (mode) of a frequency distribution is the value of the variable (or the interval) that has the largest number of observations. If the data in a sample are ranked in order of the values of the variable (highest to lowest), the *median* observation is in the middle of the ranked data.

Two distributions with the same central tendency may differ considerably in the extent to which the observations deviate around the central tendency. *Measures of dispersion* indicate the variability of the data. There are two types of measures of dispersion, absolute and rela-

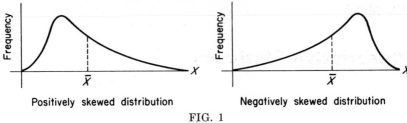

Positively skewed distribution   Negatively skewed distribution

FIG. 1

tive. The *variance*, the *standard deviation* (the square root of the variance), and the *range* are measures of absolute dispersion. The *coefficient of variation* (standard deviation divided by the arithmetic mean) and the *variance of the natural logarithm of the variable* are common measures of relative dispersion. Unlike measures of absolute dispersion or of central tendency, measures of relative dispersion are pure numbers (devoid of units).

Skewness is one measure of the shape of a distribution. A distribution is said to have *zero skewness* if it is symmetric around the arithmetic mean $(\bar{X})$ of the distribution. A distribution is *positively skewed* if the "elongated tail" is to the right (higher values). It is *negatively skewed* if the elongated tail is to the left (lower values). Positive and negative skewness are illustrated in Figure 1.

The problems in this chapter demonstrate the computation and use of frequency distributions and measures of central tendency, dispersion, and shape. To facilitate the presentation of these procedures, a useful tool, the *summation notation*, is explained in Problems 4 and 5.

Problems

1. Frequency Distribution
2. Histogram and Frequency Curve
3. Frequency Distribution
4. Summation Notation
5. Summation Notation
6. Measure of Central Tendency: Arithmetic Mean
7. Measure of Central Tendency: Median
8. Measure of Central Tendency: Mode
9. Absolute Dispersion: Variance, Standard Deviation, and Range
10. Relative Dispersion: Coefficient of Variation and Variance of the Natural Logarithm of the Variable
11. Measure of Shape: Skewness
12. Comparison of Two Frequency Distributions

Problem 1. Frequency Distribution

Q: Represent the following set of scores on an exam as a frequency distribution:

2, 7, 6, 5, 6, 2, 3, 4, 3, 2.

A: The frequency of a score is the number of times it occurs in the data. The frequency of these values is:

Value	Scoring of frequencies	Frequency
2	III	3
3	II	2
4	I	1
5	I	1
6	II	2
7	I	1
		10 = Sum

A relative frequency distribution relates the value of the variable to the relative frequency—the number of observations of that value divided by the total number of observations. The relative frequency distribution of the above values is:

Value	Relative frequency
2	0.3
3	0.2
4	0.1
5	0.1
6	0.2
7	0.1
	1.0 = Sum of relative frequencies

The sum of relative frequencies always equals 1.0.

Problem 2. Histogram and Frequency Curve

Q: Represent the following data by

   (a) a histogram,
   (b) a frequency curve.
   Data: 2, 7, 6, 5, 6, 2, 3, 4, 3, 2, 9

A: We can represent the data in interval notation.

Value (Midpoint of interval)	Interval	Frequency
2	1.50 to 2.49	3
3	2.50 to 3.49	2
4	3.50 to 4.49	1
5	4.50 to 5.49	1
6	5.50 to 6.49	2
7	6.50 to 7.49	1
8	7.50 to 8.49	0
9	8.50 to 9.49	1

   (a) A histogram (bar graph) representing these data is shown in Figure 2.

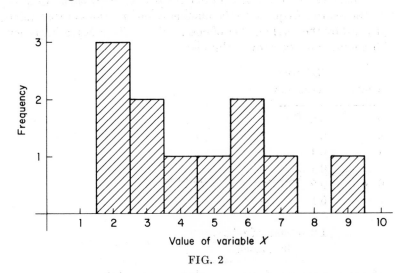

FIG. 2

(b) The frequency curve for these data is given in Figure 3.

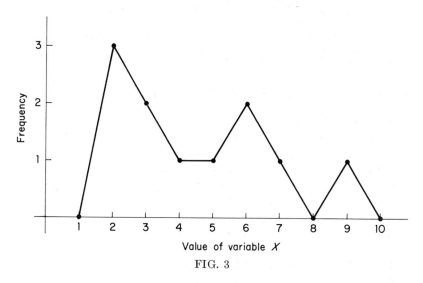

FIG. 3

Problem 3. Frequency Distribution

Q: Arrange the following set of student grade point averages (GPA), based on a 0 to 6 grading scale, as a frequency distribution and construct a frequency curve.

3.4	4.9	2.0	1.1	1.6	3.4
1.8	1.4	1.6	2.7	3.4	4.6
2.3	3.7	4.3	3.7	2.7	3.2
5.7	2.1	2.7	5.0	3.9	2.7
4.2	1.3	5.3	3.0	4.2	1.6
3.3	1.8	3.6	3.5	1.6	5.3
2.8	3.2	1.7	2.3	2.8	2.8
4.7	5.6	2.9	5.2	3.9	1.3
1.5	1.7	4.0	4.6	4.3	2.6
2.2	2.8	5.7	3.8	5.7	5.7

A: Listing the intervals first, we can then compute the interval mid-points and the absolute and relative frequencies.

Interval	Midpoint	Score	Absolute frequency	Relative frequency
1.01–1.50	1.25	Ⅲ̶Ⅱ	5	0.0833
1.51–2.00	1.75	Ⅲ̶Ⅱ ⅢⅠ	9	0.1500
2.01–2.50	2.25	ⅢⅠ	4	0.0667
2.51–3.00	2.75	Ⅲ̶Ⅱ ⅢⅡ Ⅰ	11	0.1833
3.01–3.50	3.25	Ⅲ̶Ⅱ Ⅱ	7	0.1167
3.51–4.00	3.75	Ⅲ̶Ⅱ Ⅱ	7	0.1167
4.01–4.50	4.25	ⅢⅠ	4	0.0667
4.51–5.00	4.75	Ⅲ̶Ⅱ	5	0.0833
5.01–5.50	5.25	Ⅲ	3	0.0500
5.51–6.00	5.75	Ⅲ̶Ⅱ	5	0.0833
		Sum = 60	60	1.0000

The frequency curve is shown in Figure 4.

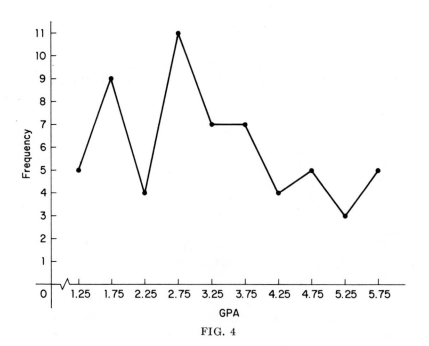

FIG. 4

Problem 4. Summation Notation

Q: Use the summation notation to compute the sum of the variable X_i, $i = 1, \ldots, 10$, given $X_1 = 2$, $X_2 = 7$, $X_3 = 6$, $X_4 = 5$, $X_5 = 6$, $X_6 = 2$, $X_7 = 3$, $X_8 = 4$, $X_9 = 3$, $X_{10} = 5$.

A: X_i represents the *i*th value of X, and $\sum\limits_{i=1}^{10} X_i$ represents the sum of the X_i's from $i = 1$ to $i = 10$. (The $\sum$ is a capital Greek sigma.) Taking the sum, we have

$$\sum_{i=1}^{10} X_i = X_1 + X_2 + \cdots + X_9 + X_{10}$$

$$= 2 + 7 + 6 + 5 + 6 + 2 + 3 + 4 + 3 + 5 = 43.$$

Problem 5. Summation Notation

Q: Use the data in Problem 4 to compute

(a) $\sum\limits_{i=1}^{N} (X_i - 1)$,   (b) $\sum\limits_{i=1}^{N} 2X_i$,   (c) $\sum\limits_{i=1}^{N} (X_i)^2$,

(d) $\sum\limits_{i=1}^{N} (X - 1)^2$,

where N is the number of observations ($N = 10$).

A: (a) $\sum\limits_{i=1}^{10} (X_i - 1) = (X_1 - 1) + (X_2 - 1) + (X_3 - 1) + \cdots$

$$+ (X_9 - 1) + (X_{10} - 1)$$

$$= X_1 + X_2 + X_3 + \cdots + X_9 + X_{10} - 10$$

$$= \sum_{i=1}^{10} X_i - \sum_{i=1}^{10} 1$$

$$= \sum_{i=1}^{10} X_i - 10$$

$$= 43 - 10 = 33.$$

(b) $\sum\limits_{i=1}^{10} 2X_i = 2X_1 + 2X_2 + 2X_3 + \cdots + 2X_9 + 2X_{10}$

$$= 2(X_1 + X_2 + X_3 + \cdots + X_9 + X_{10})$$

$$= 2\sum_{i=1}^{10} X_i = 2(43) = 86.$$

(c) $\sum\limits_{i=1}^{10} (X_i)^2 = X_1^2 + X_2^2 + X_3^2 + \cdots + X_9^2 + X_{10}^2$

$$= 2^2 + 7^2 + 6^2 + \cdots + 3^2 + 5^2.$$

$$= 213.$$

(d) $\sum\limits_{i=1}^{N} (X_i - 1)^2 = \sum\limits_{i=1}^{N} (X_i^2 - 2X_i + 1)$

$$= \sum_{i=1}^{N} X_i^2 - 2\sum_{i=1}^{N} X_i + N$$

$$= 213 - 2(43) + 10 = 137.$$

Note: The *rules of summation* for the variable X_i are: If k is a constant,

(a) $\sum\limits_{i=1}^{N} (X_i - k) = \sum\limits_{i=1}^{N} X_i - Nk,$

(b) $\sum\limits_{i=1}^{N} kX_i = k \sum\limits_{i=1}^{N} X_i,$

(c) $\sum\limits_{i=1}^{N} (X_i)^k = X_1^k + X_2^k + \cdots + X_N^k,$

(d) If Y_i is a variable, $\sum\limits_{i=1}^{N} (X_i + Y_i) = \sum\limits_{i=1}^{N} X_i + \sum\limits_{i=1}^{N} Y_i.$

Problem 6. Measure of Central Tendency: Arithmetic Mean

Q: Compute the arithmetic mean of the data in Problem 4

(a) using the simple formula,

(b) using the frequency distribution notation.

A: (a) The arithmetic mean ($\bar{X}$) is a measure of the central tendency or location of a frequency distribution. It is the sum of the values of the observations divided by the number of observations:

$$\bar{X} = \frac{\sum\limits_{i=1}^{N} X_i}{N} = \frac{43}{10} = 4.3.$$

(b) Converting the data to a frequency distribution, we have (for N observations, k intervals, and f_i observations in the interval X_i):

X_i	f_i	$f_i X_i$
2	2	4
3	2	6
4	1	4
5	2	10
6	2	12
7	1	7
	$N = 10$	43

Using frequency notation, we get

$$\sum_{i=1}^{k} f_i = N \quad \text{and} \quad \bar{X} = \frac{\sum\limits_{i=1}^{k} f_i X_i}{N},$$

$$\bar{X} = \frac{\sum\limits_{i=1}^{k} f_i X_i}{N} = \frac{2 \cdot 2 + 2 \cdot 3 + 1 \cdot 4 + 2 \cdot 5 + 2 \cdot 6 + 1 \cdot 7}{10} = \frac{43}{10}.$$

The arithmetic mean is $\bar{X} = 4.3$.

Problem 7. Measure of Central Tendency: Median

Q: Compute the median of the following sets of data.

(a) Odd number of observations:

2, 4, 5, 8, 11, 13, 14.

(b) Even number of observations:

2, 4, 5, 8, 11, 13, 14, 15.

(c)

X	f
2	5
3	6
4	8
5	4
6	3
	26 $(= N)$

(d)

X	f
2	3
3	10
4	8
5	4
6	1
	26 $(= N)$

A: The median is a measure of central tendency. It is most easily computed when the data are ranked in ascending order. The median observation in a ranked set of observations neither exceeds nor is exceeded by more than half of the observations. If there is an odd number of observations, the median is the middle observation. If there is an even number of observations, any of the numbers at or between the two middle observations can be called the median. By convention, however, the midpoint (arithmetic mean) of the two middle observations is used as the median. If N is the number of observations, the median observation is the $(N + 1)/2$-th observation.

(a) There are seven observations and when the data are ranked in ascending order, the fourth observation, 8, is the middle observation and is the median.

(b) There are eight observations and when the data are ranked in ascending order, the fourth and fifth observations (values 8 and 11) are the middle observations. The median is $(8 + 11)/2 = 9.5$.

(c) There are 26 observations and the median is the midpoint between the 13th and 14th observations. Both of these observations have the value 4. The median is 4.

(d) There are 26 observations and the median is the midpoint between the 13th and 14th observations. The 13th observation has the value 3. The 14th observation has the value 4. The median is 3.5 (the mean of 3 and 4).

Problem 8. Measure of Central Tendency: Mode

Q: Find the mode(s) of the following distributions.

(a) 2, 3, 3, 4, 5, 6

(b) 6, 2, 7, 8, 1, 2, 2, 7, 9, 3, 7

(c) 6, 2, 7, 8, 1, 2, 2, 7, 9, 3, 7, 7, 7

(d)

Variable X	Frequency
1.25	5
1.75	9
2.25	4
2.75	11
3.25	7
3.75	7
4.25	4
4.75	5
5.25	3
5.75	5
	60

A: The mode is the most frequently occurring observation in a set of observations. A distribution with one mode is called a *unimodal* distribution. A distribution with two modes is *bimodal*. Often, if an observation is the most frequent in its neighborhood, even if it is not the most frequent in the distribution, it will be referred to as a (local) mode.

(a) Construct a frequency distribution:

X	f
2	1
3	2
4	1
5	1
6	1

The mode is 3 and the distribution is unimodal.

(b) From the frequency distribution

X	f
1	1
2	3
3	1
4	0
5	0
6	1
7	3
8	1
9	1

we see that the most frequent observations are 2 and 7, both of which occur three times. The distribution is bimodal.

(c) Here the frequency distribution

X	f
1	1
2	3
3	1
4	0
5	0
6	1
7	5
8	1
9	1

indicates that the most frequent observation occurs five times and is 7. The observation 2 occurs three times and is the most frequent observation in its neighborhood. The distribution is bimodal with 7 as the mode of the distribution and 2 as a local mode.

(d) The most frequent observation is 2.75 and it is the mode. The values 1.75 and 4.75 can be viewed as local modes.

Problem 9. Absolute Dispersion: Variance, Standard Deviation and Range

Q: Compute the (a) variance, (b) standard deviation, and (c) range of the data in Problem 4. The data are 2, 7, 6, 5, 6, 2, 3, 4, 3, 5. The frequency distribution is:

X_i	f_i
2	2
3	2
4	1
5	2
6	2
7	1

A: (a) The variance is a measure of the deviation of observations from the mean. $\sum_{i=1}^{N} (X_i - \bar{X})$ would not be a satisfactory measure since

$$\sum_{i=1}^{N} (X_i - \bar{X}) = \sum_{i=1}^{N} X_i - N(\bar{X}) = 0.$$

The variance (S^2) is the sum of the *squared* deviations of the observations from the mean divided by $N - 1$:

$$S^2 = \frac{\sum_{i=1}^{N} (X_i - \bar{X})^2}{N - 1}.$$

(The variance is often written with N in the denominator. For large samples the difference is trivial. The denominator $N - 1$ is preferred when the sample variance is used in statistical analysis.)

$$\bar{X} = \frac{\sum\limits_{i=1}^{N} X_i}{N} = \frac{43}{10} = 4.3.$$

X_i	$\bar{X}$	$X_i - \bar{X}$	$(X_i - \bar{X})^2$
2	4.3	−2.3	5.29
2	4.3	−2.3	5.29
3	4.3	−1.3	1.69
3	4.3	−1.3	1.69
4	4.3	−0.3	0.09
5	4.3	0.7	0.49
5	4.3	0.7	0.49
6	4.3	1.7	2.89
6	4.3	1.7	2.89
7	4.3	2.7	7.29

Sum: 0.0   28.10

$$S^2 = \frac{\sum\limits_{i=1}^{10} (X_i - \bar{X})^2}{N - 1} = \frac{28.10}{10 - 1} = 3.12.$$

Using frequency distribution notation, we have

$$S^2 = \frac{\sum\limits_{i=1}^{k} f_i(X_i - \bar{X})}{N - 1},$$

where k is the number of intervals and

$$\sum_{i=1}^{k} f_i = N.$$

X_i	f_i	$\bar{X}$	$X_i - \bar{X}$	$(X_i - \bar{X})^2$	$f_i(X_i - \bar{X})^2$
2	2	4.3	-2.3	5.29	10.58
3	2	4.3	-1.3	1.69	3.38
4	1	4.3	-0.3	0.09	0.09
5	2	4.3	0.7	0.49	0.98
6	2	4.3	1.7	2.89	5.78
7	1	4.3	2.7	7.29	7.29
Sum:			0.0		28.10

$$S^2 = \frac{\sum_{i=1}^{6} f_i(X_i - \bar{X})^2}{N - 1} = \frac{28.10}{10 - 1} = 3.12.$$

Note: There is an alternative computational procedure which is simpler for large samples:†

$$\sum f(X - \bar{X})^2 = \sum fX^2 - \sum f(2X)\bar{X} + \sum f\bar{X}^2$$

$$= \sum fX^2 - N\bar{X}^2,$$

$$S^2 = \frac{\sum fX^2}{N - 1} - \frac{(\bar{X})^2 N}{N - 1}.$$

† $\sum_{i=1}^{k}$ is replaced by $\sum$, and the subscript i notation is suppressed for f_i znd X_i.

X_i	f_i	f_iX_i	X_i^2	$f_iX_i^2$
2	2	4	4	8
3	2	6	9	18
4	1	4	16	16
5	2	10	25	50
6	2	12	36	72
7	1	7	49	49
—	—	—		—
	10	43		213

$$S^2 = \frac{\sum f_iX_i^2}{N-1} - \left(\frac{\sum f_iX_i}{N}\right)^2 \left(\frac{N}{N-1}\right)$$

$$= \frac{213}{9} - (4.3)^2 \left(\frac{10}{9}\right) = \frac{213 - 184.9}{9} = \frac{28.10}{9} = 3.12.$$

For large samples $N - 1 \approx N$ and the formula for variance can be written as

$$S^2 = \frac{\sum\limits_{i=1}^{N} f_iX_i^2}{N} - \bar{X}^2.$$

(b) The standard deviation (S) is the positive square root of the variance:

$$S = \sqrt{S^2} = \sqrt{3.12} = 1.77.$$

The units of the standard deviation are the same as the units of the data. For example, if the units of the data are dollars, the units of the standard deviation are also dollars, and the units of the variance of income are "dollars squared."

(c) The range of a set of data is the difference between the highest and the lowest values: $7 - 2 = 5$. The range of the data is 5 units.

Problem 10. Relative Dispersion: Coefficient of Variation and Variance of the Natural Logarithm of the Variable

Q: Use the data in the previous problem to compute

(a) the coefficient of variation,
(b) the variance of the natural logarithm of the variable.

A: Measures of relative dispersion are pure numbers; they are devoid of units.

(a) The coefficient of variation is the standard deviation divided by the mean:

$$\text{CV}\,(X) = \frac{S(X)}{\bar{X}} = \frac{1.77}{4.3} = 0.41.$$

(b) The variance of the natural logarithm (ln) of the variable X is

$$S^2(\ln X) = \frac{\sum_{i=1}^{k} [\ln X_i - \overline{(\ln X)}]^2}{N - 1},$$

where $\overline{(\ln X)}$ is the mean of the natural log of the variable X.†

† The natural log of the geometric mean of X is

$$\ln\,[\text{GM}\,(X)] = \ln\,(\sqrt[N]{X_1 \cdots X_N}) = \frac{1}{N}\,[\ln\,(X_1 \cdots X_N)]$$

$$= \frac{1}{N}\,(\ln X_1 + \cdots + \ln X_N) = \overline{(\ln X)}.$$

X_i	f_i	$\ln X_i$	$f_i(\ln X_i)$	$\ln X - \overline{\ln X}$	$(\ln X - \overline{\ln X})^2$	$f(\ln X - \overline{\ln X})^2$
2	2	0.69	1.38	-0.68	0.46	0.92
3	2	1.10	2.20	-0.27	0.07	0.15
4	1	1.39	1.39	0.02	0.00	0.00
5	2	1.61	3.22	0.24	0.06	0.12
6	2	1.79	3.58	0.42	0.18	0.35
7	1	1.95	1.95	0.58	0.34	0.34
			13.72			1.88

$$\overline{\ln X} = \frac{\sum f(\ln X)}{N} = \frac{13.72}{10} = 1.37,$$

$$S^2(\ln X) = \frac{\sum\limits_{i=1}^{6} f_i(\ln X_i - \ln X)^2}{N-1} = \frac{1.88}{9} = 0.21.$$

The variance of the natural log of X is 0.21.

Problem 11. Measure of Shape: Skewness

Q: Are any of the following three frequency distributions skewed?

(a) X_i	f_i	(b) X_i	f_i	(c) X_i	f_i
2	2	2	1	2	1
3	2	3	2	3	1
4	2	4	2	4	2
5	2	5	2	5	2
6	1	6	2	6	2
7	1	7	1	7	2

A: Skewness refers to the lack of symmetry of a distribution around the mean. A symmetric distribution is not skewed. In a symmetric distribution the mean equals the median. Also, in a symmetric distribution the third moment of the variable around the mean equals zero, that is,

$$\sum_{i=1}^{k} f_i(X_i - \bar{X})^3 = 0.$$

A distribution is said to be positively skewed if its right tail is longer than its left tail. For a positively skewed distribution the arithmetic mean is larger than the median, and the third moment of the variable around the mean is positive, that is

$$\sum_{i=1}^{k} f_i(X_i - \bar{X})^3 > 0.$$

In a negatively skewed distribution the elongated tail is to the left, the arithmetic mean is less than the median, and the third moment around the mean is negative:

$$\sum_{i=1}^{k} f_i(X_i - \bar{X})^3 < 0.$$

There are several measures of skewness. For example,

1. $Sk\ (X) = \dfrac{\bar{X} - \text{Med}}{S}$,

2. $Sk\ (X) = \dfrac{\sum f_i(X_i - \bar{X})^3}{N - 1} = \mu_3,$

3. $Sk\ (X) = \dfrac{(\mu_3)^2}{(S^2)^3}$ or $\sqrt[6]{\dfrac{(\mu_3)^2}{(S^2)^3}},$

where

$$\mu_3 = \frac{\sum f_i(X_i - \bar{X})^3}{N - 1}$$

and S^2 is the variance.

The first and third measures are pure numbers, and they measure skewness relative to the dispersion in the distribution.

(a) Using the first frequency distribution, we have:

X_i	f_i	f_iX_i	$X_i - \bar{X}$	$(X_i - \bar{X})^2$	$(X_i - \bar{X})^3$	$f_i(X_i - \bar{X})^3$
2	2	4	-2.1	4.41	-9.26	-18.52
3	2	6	-1.1	1.21	-1.33	$- 2.66$
4	2	8	-0.1	0.01	-0.00	$- 0.00$
5	2	10	0.9	0.81	0.73	1.46
6	1	6	1.9	3.61	6.86	6.86
7	1	7	2.9	8.41	24.39	24.39
Sum:		41				11.45

$$\bar{X} = \frac{\sum f_iX_i}{N} = \frac{41}{10} = 4.1,$$

$$\text{Sk } (X) = \frac{\sum f_i(X_i - \bar{X})^3}{N - 1} = \frac{11.45}{9} = 1.27,$$

Med = 4.

The measure of skewness is positive, and the mean is larger than the median. The distribution is positively skewed (see Figure 5).

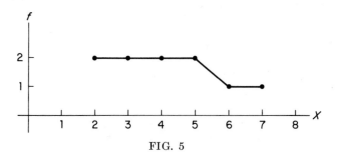

FIG. 5

(b) For the second distribution:

X_i	f_i	f_iX_i	$X_i - \bar{X}$	$(X_i - \bar{X})^2$	$(X_i - \bar{X})^3$	$f_i(X_i - \bar{X})^3$
2	1	2	−2.5	6.25	−15.63	−15.63
3	2	6	−1.5	2.25	− 3.38	− 6.76
4	2	8	−0.5	0.25	− 0.13	− 0.26
5	2	10	0.5	0.25	0.13	0.26
6	2	12	1.5	2.25	3.38	6.76
7	1	7	2.5	6.25	15.63	15.63
Sum:		45				0.00

$$\bar{X} = \frac{45}{10} = 4.5,$$

$$\text{Sk } (X) = \frac{\sum f_i(X_i - \bar{X})^3}{N - 1} = \frac{0}{9} = 0.$$

Skewness is zero; the distribution is symmetric around the mean (see Figure 6). The median is in the interval between 4 and 5. If we use the midpoint of the interval to represent the interval, the mean and median are equal.

(c) For the third distribution:

X_i	f_i	f_iX_i	$X_i - \bar{X}$	$(X_i - \bar{X})^2$	$(X_i - \bar{X})^3$	$f_i(X_i - \bar{X})^3$
2	1	2	−2.9	8.41	−24.39	−24.39
3	1	3	−1.9	3.61	− 6.86	− 6.86
4	2	8	−0.9	0.81	− 0.73	− 1.46
5	2	10	0.1	0.01	0.00	0.00
6	2	12	1.1	1.21	1.33	2.66
7	2	14	2.1	4.41	9.26	18.52
Sum:		49				−11.53

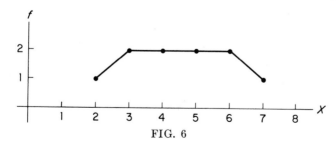

FIG. 6

$$\bar{X} = \frac{\sum f_i X_i}{N} = \frac{49}{10} = 4.9,$$

$$\text{Sk } (X) = \frac{\sum f_i (X_i - \bar{X})^3}{N - 1} = \frac{-11.53}{9} = -1.28,$$

$$\text{Med} = 5.$$

The measure of skewness is negative, and the mean is less than the median. The distribution is negatively skewed.

Note: When a distribution has one mode (unimodal), there is a relationship between the mode (Mo) and the mean ($\bar{X}$) or median (Med).

1. For a positively skewed distribution, Mo $<$ Med $< \bar{X}$, as shown in Figure 7.

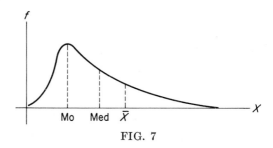

FIG. 7

2. For a symmetric distribution, Mo $=$ Med $= \bar{X}$, as shown in Figure 8.

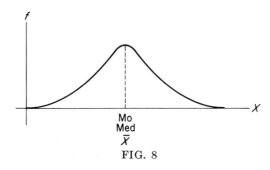

FIG. 8

3. For a negatively skewed distribution, Mo > Med > $\bar{X}$, as shown in Figure 9.

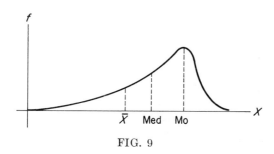

FIG. 9

Problem 12. Comparison of Two Frequency Distributions

Q: Frequency distributions A and B represent the hourly wage rates of male (A) and female (B) workers in a company. Compare the two frequency distributions using the tools of descriptive statistics developed in this chapter.

A (Males)		B (Females)	
X_i	f_i	Y_i	f_i
1.75	3	1.75	2
2.00	3	2.00	4
2.25	5	2.25	5
2.50	7	2.50	8
2.75	7	2.75	7
3.00	10	3.00	5
3.25	8	3.25	4
3.50	6	3.50	3
3.75	4	3.75	2
4.00	3		—
4.25	2		40
4.50	2		
	—		
	60		

Here X = hourly wage rate for males, Y = hourly wage rate for females, and the values are interval midpoints.

A: For males:

X_i	f_i	f_iX_i	$X_i - \bar{X}$	$(X_i - \bar{X})^2$	$f_i(X_i - \bar{X})^2$	$(X_i - \bar{X})^3$	$f_i(X_i - \bar{X})^3$
1.75	3	5.25	-1.26	1.59	4.77	-2.00	-6.00
2.00	3	6.00	-1.01	1.02	3.06	-1.03	-3.09
2.25	5	11.25	-0.76	0.58	2.90	-0.44	-2.20
2.50	7	17.50	-0.51	0.26	1.82	-0.13	-0.91
2.75	7	19.25	-0.26	0.07	0.49	-0.02	-0.14
3.00	10	30.00	-0.01	0.00	0.00	-0.00	-0.00
3.25	8	26.00	0.24	0.06	0.48	0.01	0.08
3.50	6	21.00	0.49	0.24	1.44	0.12	0.72
3.75	4	15.00	0.74	0.55	2.20	0.41	1.64
4.00	3	12.00	0.99	0.98	2.94	0.97	2.91
4.25	2	8.50	1.24	1.54	3.08	1.91	3.82
4.50	2	9.00	1.49	2.22	4.44	3.31	6.62
	60	180.75			27.62		3.45

$$\bar{X} = \frac{\sum f_iX_i}{N} = \frac{180.75}{60} = 3.01.$$

For females:

Y_i	f_i	f_iY_i	$Y_i - \bar{Y}$	$(Y_i - \bar{Y})^2$	$f_i(Y_i - \bar{Y})^2$	$(Y_i - \bar{Y})^3$	$f_i(Y_i - \bar{Y})^3$
1.75	2	3.50	-0.95	0.90	1.80	-0.86	-1.72
2.00	4	8.00	-0.70	0.49	1.96	-0.34	-1.36
2.25	5	11.25	-0.45	0.20	1.00	-0.09	-0.45
2.50	8	20.00	-0.20	0.04	0.32	-0.01	-0.08
2.75	7	19.25	0.05	0.00	0.02	0.00	0.00
3.00	5	15.00	0.30	0.09	0.45	0.03	0.15
3.25	4	13.00	0.55	0.30	1.20	0.17	0.68
3.50	3	10.50	0.80	0.64	1.92	0.51	1.53
3.75	2	7.50	1.05	1.10	2.20	1.16	2.32
	40	108.00			10.87		1.07

$$\bar{Y} = \frac{\sum\limits_{i=1}^{N} f_iY_i}{N} = \frac{108}{40} = 2.7.$$

1. *Measures of Central Tendency*

(a) Arithmetic mean: The arithmetic mean hourly wage rate is greater for men ($\bar{X}$ = $3.01) than for women ($\bar{Y}$ = $2.70).
(b) Median: The median observation is the $(N + 1)/2$-th observation.
    For men, $(60 + 1)/2 = 30.5$. Therefore the median is between the 30th and 31st observations. The median for men is in the interval $2.875–$3.125 (the midpoint of the interval is $3.00).
    For women, $(40 + 1)/2 = 20.5$, and the median wage is in the interval $2.625–$2.875 (the midpoint of the interval is $2.75).
(c) Mode: The modal wage is in the interval with the midpoint $3.00 for men and $2.50 for women.

The three measures of central tendency indicate a higher level of hourly wage rates for men than for women.

2. *Absolute and Relative Dispersion*

(a) Range: The range of the data is $4.50 − $1.75 = $2.75 for men and $3.75 − $1.75 = $2.00 for women.
(b) Variance: For men

$$S^2(X) = \frac{\sum\limits_{i=1}^{N} f_i(X_i - \bar{X})^2}{N - 1} = \frac{27.62}{59} = 0.47.$$

For women

$$S^2(Y) = \frac{\sum\limits_{i=1}^{N} f_i(Y_i - \bar{Y})^2}{N - 1} = \frac{10.87}{39} = 0.28.$$

(c) Coefficient of variation:

$$CV\ (X) = \frac{S(X)}{\bar{X}} = \frac{\sqrt{0.47}}{3.01} = \frac{0.69}{3.01} = 0.23,$$

$$CV\ (Y) = \frac{S(Y)}{\bar{Y}} = \frac{\sqrt{0.28}}{2.70} = \frac{0.53}{2.70} = 0.20.$$

The measures of absolute dispersion (range and variance) and relative dispersion (coefficient of variation) indicate greater inequality in hourly wage rates for men than for women.

3. *Skewness*

(a) Absolute skewness:

$$\mathrm{Sk}\ (X) = \frac{\sum_{i=1}^{N} f_i (X_i - \bar{X})^3}{N - 1} = \frac{+3.45}{59} = 0.058,$$

$$\mathrm{Sk}\ (Y) = \frac{\sum_{i=1}^{N} f_i (Y_i - \bar{Y})^3}{N - 1} = \frac{+1.07}{39} = 0.027.$$

(b) Relative skewness:

$$\mathrm{Sk}\ (X) = \frac{(\mu_3)^2}{(S^2)^3} = \frac{(0.058)^2}{(0.47)^3} = \frac{0.00336}{0.10382} = 0.0323,$$

$$\mathrm{Sk}\ (Y) = \frac{(\mu_3)^2}{(S^2)^3} = \frac{(0.027)^2}{(0.28)^3} = \frac{0.00073}{0.02195} = 0.0333.$$

Both distributions are positively skewed. The male wage distribution has a larger absolute skewness, but when standardized by the variance there is almost no difference in skewness in hourly wage rates.

In sum, the male wage distribution has a higher level and a greater dispersion than the female distribution. Both distributions are positively skewed, with the absolute skewness greater for males than for females.

If these data were obtained from random sampling from a population of male and female wage rates, we would want to know whether the observed differences represented differences in the distribution of wages in the two populations or whether the observed differences could be explained by sampling fluctuations. The procedures developed in Chapters 3 and 4 can answer this question.

Problem 13. Covariance and Correlation

Q: Two variables X and Y are measured for a set of N observations. What are the covariance and correlation coefficient for each set of data?

(a) X = family income in thousands of dollars, Y = years of schooling of the head of the household.

Family	X	Y
1	10	12
2	9	12
3	13	15
4	12	14
5	16	15
6	18	16

(b) X = family income in thousands of dollars, Y = number of children.

Family	X	Y
1	10	2
2	9	3
3	13	1
4	12	1
5	16	1
6	18	0

(c) X = dollar value of net assets of husband, Y = dollar value of net assets of wife.

Family	X	Y
1	6	−8.00
2	5	−8.66
3	6	8.00
4	−6	−8.00
5	−5	8.66
6	−6	8.00
7	5	8.66
8	−5	−8.66

A: The covariance and the correlation coefficient are two measures of the joint variation of two variables. If one variable X tends to be higher for higher values of the second variable Y, variables X and Y are said to be *positively correlated* or to have a *positive covariance*. If X tends to be higher when Y is lower, the variables are said to be *negatively correlated* or to have a *negative covariance*. If X tends to be neither higher nor lower for higher values of Y, the correlation coefficient and the covariance are zero. A nonzero correlation (covariance) implies an association between X and Y. A zero correlation (covariance) implies the absence of a linear relation (association) between X and Y, but does not indicate whether a nonlinear relation exists or whether the two variables are independent. If two variables X and Y are independent, there is no relation (linear or nonlinear) between the value of X and Y.

The covariance of X and Y is written as

$$\text{Cov}(X,Y) = \frac{\sum\limits_{i=1}^{N} (X_i - \bar{X})(Y_i - \bar{Y})}{N}.$$

This shows the joint variation of X_i and Y_i. Since

$$\sum_{i=1}^{N} (X_i - \bar{X})(Y_i - \bar{Y}) = \sum_{i=1}^{N} X_i Y_i - \bar{X}\sum_{i=1}^{N} Y_i - \bar{Y}\sum_{i=1}^{N} X_i + \sum_{i=1}^{N} \bar{X}\bar{Y}$$

$$= \sum_{i=1}^{N} X_i Y_i - N\bar{X}\bar{Y},$$

it follows that

$$\text{Cov}(X,Y) = \frac{\sum\limits_{i=1}^{N} X_i Y_i}{N} - \bar{X}\bar{Y}.$$

If higher values of X are associated with higher values of Y, the mean of the product of X and Y is larger than the product of the means and Cov $(X,Y) > 0$. If higher values of X are associated with lower values of Y, the mean of the product of X and Y is less than the product of the means, and Cov $(X,Y) < 0$. The covariance is in the units of X and Y.

The correlation coefficient $[R(X,Y)]$ is referred to as the covariance in standard units. It is the covariance divided by the product of the standard deviations of X and Y (where N rather than $N - 1$ is in the denominator of the variance):

$$R(X,Y) = \frac{\text{Cov }(X,Y)}{S(X)S(Y)} .$$

The correlation coefficient is a pure number and is in the interval bounded by -1.0 and $+1.0$, that is, $-1.0 \le R(X,Y) \le +1.0$.

(a)

X_i	Y_i	$X_i - \bar{X}$	$(X_i - \bar{X})^2$	$Y_i - \bar{Y}$	$(Y_i - \bar{Y})^2$	$[(X_i - \bar{X}) \times (Y_i - \bar{Y})]$
10	12	-3	9	-2	4	6
9	12	-4	16	-2	4	8
13	15	0	0	1	1	0
12	14	-1	1	0	0	0
16	15	3	9	1	1	3
18	16	5	25	2	4	10
Sum: 78	84	0	60	0	14	27

$$\bar{X} = \frac{\sum\limits_{i=1}^{N} X_i}{N} = \frac{78}{6} = 13,$$

$$S^2(X) = \sum_{i=1}^{N} \frac{(X_i - \bar{X})^2}{N} = \frac{60}{6} = 10, \qquad S(X) = 3.16,$$

$$\bar{Y} = \frac{\sum\limits_{i=1}^{N} Y_i}{N} = \frac{84}{6} = 14,$$

$$S^2(Y) = \frac{\sum\limits_{i=1}^{N} (Y_i - \bar{Y})^2}{N} = \frac{14}{6} = 2.33, \qquad S(Y) = 1.53,$$

$$\text{Cov}(X,Y) = \frac{\sum\limits_{i=1}^{N} (X_i - \bar{X})(Y_i - \bar{Y})}{N} = \frac{27}{6} = 4.50,$$

$$R(X,Y) = \frac{\text{Cov}(X,Y)}{S(X)S(Y)} = \frac{4.50}{(3.16)(1.53)} = 0.93.$$

The covariance between X and Y is 4.50, and the correlation coefficient is 0.93. There is a positive relation (association) between X and Y, as illustrated in Figure 10.

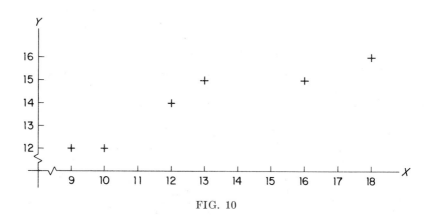

FIG. 10

(b)

X_i	Y_i	$X_i - \bar{X}$	$(X_i - \bar{X})^2$	$Y_i - \bar{Y}$	$(Y_i - \bar{Y})^2$	$[(X_i - \bar{X}) \times (Y_i - \bar{Y})]$
10	2	-3	9	0.67	0.45	-2.01
9	3	-4	16	1.67	2.79	-6.68
13	1	0	0	-0.33	0.11	0.00
12	1	-1	1	-0.33	0.11	0.33
16	1	3	9	-0.33	0.11	-0.99
18	0	5	25	-1.33	1.77	-6.65
Sum: 78	8	0	60	0.00	5.34	-16.00

$$\bar{X} = \frac{\sum\limits_{i=1}^{N} X_i}{N} = \frac{78}{6} = 13,$$

$$S^2(X) = \frac{\sum\limits_{i=1}^{N} (X_i - \bar{X})^2}{N} = \frac{60}{6} = 10, \quad S(X) = 3.16,$$

$$\bar{Y} = \frac{\sum\limits_{i=1}^{N} Y_i}{N} = \frac{8}{6} = 1.33,$$

$$S^2(Y) = \frac{\sum\limits_{i=1}^{N} (Y_i - \bar{Y})^2}{N} = \frac{5.34}{6} = 0.89, \quad S(Y) = 0.94,$$

$$\text{Cov}(X,Y) = \frac{\sum\limits_{i=1}^{N} (X_i - \bar{X})(Y_i - \bar{Y})}{N} = \frac{-16}{6} = -2.67,$$

$$R(X,Y) = \frac{\text{Cov}(X,Y)}{S(X)S(Y)} = \frac{-2.67}{(3.16)(0.94)} = -0.9.$$

The covariance is -2.67, and the correlation coefficient is -0.9. There is a negative relation (association) between X and Y, as shown in Figure 11.

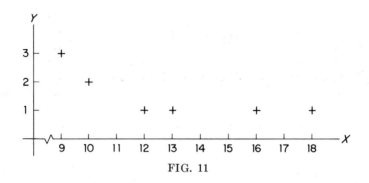

FIG. 11

(c)

X_i	Y_i	$X_i - \bar{X}$	$(X_i - \bar{X})^2$	$Y_i - \bar{Y}$	$(Y_i - \bar{Y})^2$	$[(X_i - \bar{X})$ $\times (Y_i - \bar{Y})]$
6	-8.00	6	36	-8.00	64	-48.00
5	-8.66	5	25	-8.66	75	-43.30
6	8.00	6	36	8.00	64	48.00
-6	-8.00	-6	36	-8.00	64	48.00
-5	8.66	-5	25	8.66	75	-43.30
-6	8.00	-6	36	8.00	64	-48.00
5	8.66	5	25	8.66	75	43.30
-5	-8.66	-5	25	-8.66	75	43.30
—	—	—	—			
Sum: 0	0.00	0	244	0.00	556	0.00

$$\bar{X} = \frac{\sum\limits_{i=1}^{N} X_i}{N} = \frac{0}{8} = 0,$$

$$S^2(X) = \frac{\sum\limits_{i=1}^{N} (X_i - \bar{X})^2}{N} = \frac{244}{8} = 30.5, \qquad S(X) = 5.5,$$

$$\bar{Y} = \frac{\sum\limits_{i=1}^{N} Y_i}{N} = \frac{0}{8} = 0,$$

$$S^2(Y) = \frac{\sum\limits_{i=1}^{N} (Y_i - \bar{Y})^2}{N} = \frac{556}{8} = 69.5, \qquad S(Y) = 8.3,$$

$$\text{Cov } (X,Y) = \frac{\sum\limits_{i=1}^{N} (X_i - \bar{X})(Y_i - \bar{Y})}{N} = \frac{0}{8} = 0,$$

$$R(X,Y) = \frac{\text{Cov } (X,Y)}{S(X)S(Y)} = \frac{0}{(5.5)(8.3)} = 0.$$

The covariance and the correlation coefficient are both equal to zero. This means that there is no linear relation between X and Y. However, an examination of the graph of X,Y indicates that X and Y are not independent but rather form a circle, as shown in Figure 12. [Actually, the data were obtained from the equation of a circle, $X^2 + Y^2 = 100$.]

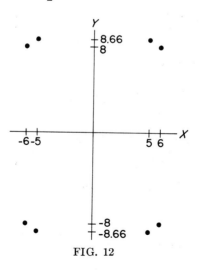

FIG. 12

Problem 14. Mean and Variance of a Sum of Distributions

Q: Suppose family income (Y) is defined as the sum of the husband's income (X_1) and the wife's income (X_2). If we know the distribution of the husband's and wife's incomes, what is (a) the mean and (b) the variance of family income? The data are as follows (income is in thousands of dollars):

Family	X_1	X_2
1	6	3
2	7	3
3	9	4
4	10	11
5	12	8

A: (a) If $Y_i = X_{1,i} + X_{2,i}$ for the ith family, summing across all N families, we have

$$\sum_{i=1}^{N} Y_i = \sum_{i=1}^{N} (X_{1,i} + X_{2,i}) = \sum_{i=1}^{N} X_{1i} + \sum_{i=1}^{N} X_{2i}$$

and dividing both sides of the equation by N,

$$\boxed{\bar{Y} = \bar{X}_1 + \bar{X}_2.}$$

Using the data we get

$$\bar{X}_1 = \frac{6 + 7 + 9 + 10 + 12}{5} = \frac{44}{5} = 8.8,$$

$$\bar{X}_2 = \frac{3 + 3 + 4 + 11 + 8}{5} = \frac{29}{5} = 5.8,$$

$$\bar{Y} = \frac{(6 + 3) + (7 + 3) + (9 + 4) + (10 + 11) + (12 + 8)}{5}$$

$$= \frac{73}{5} = 14.6.$$

Thus we can verify that

$$\bar{Y} = \bar{X}_1 + \bar{X}_2 = 8.8 + 5.8 = 14.6.$$

(b) If $Y_i = X_{1,i} + X_{2,i}$, for the ith family, subtracting the mean from both sides of the equation, we get

$$Y_i - \bar{Y} = (X_{1,i} + X_{2,i}) - (\bar{X}_1 + \bar{X}_2)$$

$$= (X_{1,i} - \bar{X}_1) + (X_{2,i} - \bar{X}_2).$$

Squaring both sides of the equation then gives

$$(Y_i - \bar{Y})^2 = [(X_{1,i} - \bar{X}_1) + (X_{2,i} - \bar{X}_2)]^2,$$

and summing across all the N observations yields

$$\sum_{i=1}^{N} (Y_i - \bar{Y})^2 = \sum_{i=1}^{N} [(X_{1,i} - \bar{X}_1) + (X_{2,i} - \bar{X}_2)]^2$$

$$= \sum_{i=1}^{N} [(X_{1,i} - \bar{X}_1)^2 + 2(X_{1,i} - \bar{X}_1)$$

$$\times (X_{2,i} - \bar{X}_2) + (X_{2,i} - \bar{X}_2)^2]$$

$$= \sum_{i=1}^{N} (X_{1,i} - \bar{X}_1)^2 + \sum_{i=1}^{N} (X_{2,i} - \bar{X}_2)^2$$

$$+ 2 \sum_{i=1}^{N} (X_{1,i} - \bar{X}_i)(X_{2,i} - \bar{X}_2).$$

Dividing both sides by N, we get

$$S^2(Y) = S^2(X_1) + S^2(X_2) + 2 \, \text{Cov} \, (X_1, X_2),$$

where Cov (X_1, X_2) is the covariance of X_1 and X_2.

X_1	X_2	$X_1 - \bar{X}_1$	$(X_1 - \bar{X}_1)^2$	$X_2 - \bar{X}_2$	$(X_2 - \bar{X}_2)^2$	$[(X_1 - \bar{X}_1) \times (X_2 - \bar{X}_2)]$
6	3	−2.8	7.84	−2.8	7.84	7.84
7	3	−1.8	3.24	−2.8	7.84	5.04
9	4	0.2	0.04	−1.8	3.24	−0.36
10	11	1.2	1.44	5.2	27.04	6.24
12	8	3.2	10.24	2.2	4.84	7.04
Sum: 44	29	0.0	22.80	0.0	50.80	25.80

$$\bar{X}_1 = \frac{\sum X_1}{N} = \frac{44}{5} = 8.8,$$

$$\bar{X}_2 = \frac{\sum X_2}{N} = \frac{29}{5} = 5.8,$$

$$S^2(X_1) = \frac{\sum (X_1 - \bar{X}_1)^2}{N} = \frac{22.8}{5} = 4.56,$$

$$S^2(X_2) = \frac{\sum (X_2 - \bar{X}_2)^2}{N} = \frac{50.80}{5} = 10.16,$$

$$\text{Cov } (X_1, X_2) = \frac{\sum (X_1 - \bar{X}_1)(X_2 - \bar{X}_2)}{N} = \frac{25.80}{5} = 5.16.$$

Letting

$$Z = S^2(X_1) + S^2(X_2) + 2 \text{ Cov } (X_1, X_2),$$

we have

$$Z = 4.56 + 10.16 + 2(5.16) = 25.04.$$

Directly computing the variance of Y, since $Y = X_1 + X_2$, we get:

Family	Y	$Y - \bar{Y}$	$(Y - \bar{Y})^2$
1	9	−5.6	31.36
2	10	−4.6	21.16
3	13	−1.6	2.56
4	21	6.4	40.96
5	20	5.4	29.16
	73	0.0	125.20

$$\bar{Y} = \frac{\sum Y}{N} = \frac{73}{5} = 14.6,$$

$$S^2(Y) = \frac{\sum (Y - \bar{Y})^2}{N} = \frac{125.20}{5} = 25.04.$$

Thus we have verified that if $Y_i = X_{1,i} + X_{2,i}$, then

$$S^2(Y) = S^2(X_1) + S^2(X_2) + 2 \text{ Cov } (X_1, X_2).$$

Note: The following are general rules for the mean and variance of a sum of distributions.

(a) If $Y_i = X_{1,i} + X_{2,i}$, then $\bar{Y} = \bar{X}_1 + \bar{X}_2$, and

$$S^2(Y) = S^2(X_1) + S^2(X_2) + 2 \text{ Cov } (X_1, X_2).$$

(b) If $Y_i = X_{1,i} - X_{2,i}$, then $\bar{Y} = \bar{X}_1 - \bar{X}_2$, and

$$S^2(Y) = S^2(X_1) + S^2(X_2) - 2 \text{ Cov } (X_1, X_2).$$

(c) If $Y_i = X_{1,i} + X_{2,i} + X_{3,i}$, then

$$\bar{Y} = \bar{X}_1 + \bar{X}_2 + \bar{X}_3 = \sum_{j=1}^{3} \bar{X}_j,$$

and

$$S^2(Y) = S^2(X_1) + S^2(X_2) + S^2(X_3) + 2 \text{ Cov } (X_1,X_3)$$
$$+ 2 \text{ Cov } (X_2,X_3) + 2 \text{ Cov } (X_1,X_2)$$
$$= \sum_{j=1}^{3} S^2(X_j) + 2 \sum_{j<k}^{3} \text{ Cov } (X_j,X_k)$$

for $j = 1,\ldots, 3$ and $k = 1,\ldots, 3$.

(d) If $Y_i = \sum_{j=1}^{N} X_{ji}$, then

$$\bar{Y} = \sum_{j=1}^{N} \bar{X}_j,$$

$$S^2(Y) = \sum_{j=1}^{N} S^2(X_j) + 2 \sum_{j<k}^{N} \text{ Cov } (X_j,X_k)$$

for $j = 1,\ldots, N$, and $k = 1,\ldots, N$.

(e) If $Y_i = (X_{1,i})(X_{2,i})$, then

$$\bar{Y} = \sum (X_1X_2) = \bar{X}_1\bar{X}_2 + \text{ Cov } (X_1,X_2).$$

From the definition of the covariance,

$$\text{Cov } (X_1,X_2) = \frac{\sum (X_1 - \bar{X}_1)(X_2 - \bar{X}_2)}{N}$$

$$= \frac{\sum (X_1X_2)}{N} - \bar{X}_1\bar{X}_2.$$

Chapter 2

Probability and Random Variables

This chapter presents procedures for solving problems in probability. Probability uses the principle of deduction; on the basis of assumptions about a population a statement is made about the probable composition of a sample. The chapter is composed of three parts: counting techniques, probability of an event, and random variables.

Counting Techniques (Problems 1 through 14)

Counting techniques provide answers to questions about the possible ways of arranging sets of objects. For example, these techniques are used to answer the question "How many committees of three members can be formed from a group of seven members?" Counting techniques are important for their own sake. They are also important as input into statements about the probability of events. For instance, the question "What is the probability that an ace will be the first card selected from a deck of 52 playing cards?" is answered by using counting techniques to find both the number of outcomes that are aces and the total number of outcomes. The ratio of the former number of outcomes to the latter number of outcomes is the probability asked for in the question.

The two basic concepts in counting techniques are *permutations* and *combinations*. The number of permutations of a set of objects is the number of arrangements of these objects in a definite order. For example, the set of two objects, A and B, can be arranged in two ways, AB and BA. Thus there are two permutations for this set. The number of combinations of a set of objects is the number of arrangements of these objects *without* concern for the order of the objects. For example, the set of two objects, A and B, can be combined in only one way if order does not matter, AB or BA. Thus there is only one combination for this set.

Probability of an Event (Problems 15 through 26)

The probability of an event A is the relative frequency with which the event A occurs. That is, if all possible outcomes are equally likely to occur, the probability of event A is the number of possible occurrences of event A divided by the total number of possible outcomes. $P(A)$ is used to designate the probability of the occurrence of event A. A probability can never be negative or greater than unity. Thus $0 \leq P(A) \leq 1$. $P(A) = 0$ means that A cannot occur. $P(A) = 1$ means that A must occur. $P(A) = 0.4$ means that event A will occur in 40 percent of the situations if an infinite number of experiments are conducted.

A probability statement can be represented graphically by a *Venn diagram* (Figure 1). The total area in the rectangle (regions 1 and 2)

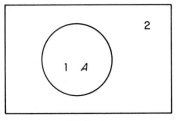

FIG. 1

represents all possible outcomes. The area in region 1 represents outcome A. The ratio of the area in region 1 to the total area in the rectangle is the probability of event A. The total area in a Venn diagram is defined to be unity, so region 1 is the probability of event A.

Union and intersection of events

The event $A \cup B$ (read "A union B") has occurred if *any one* of the three following situations is true:

A	B
occurred	occurred
occurred	did not occur
did not occur	occurred

The event $A \cap B$ (read "A intersection B") has occurred if *both* A and B have occurred:

A	B
occurred	occurred

In probability theory there is a very important formula, the *additive rule*, which states that for any two events A and B,

$$P(A \cup B) = P(A) + P(B) - P(A \cap B).$$

In words, the probability that either event A or event B occurs is equal to the probability of event A plus the probability of event B, minus the probability of both occurring, to avoid double counting. The Venn diagram in Figure 2 may help to explain this formula. In this figure

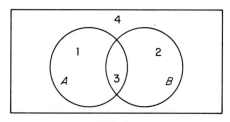

FIG. 2

region 1 is A and *not B*; region 2 is B and *not A*; region 3 is *both A* and *B*; and region 4 is *neither A nor B*. Therefore $P(A \cup B)$ may be represented by region 1 + region 2 + region 3. $P(A)$ may be represented by region 1 + region 3. $P(B)$ may be represented by region 2 + region 3. $P(A \cap B)$ may be represented by region 3. Then since

$$P(A) + P(B) - \left(P(A \cap B) \right)$$

$$= (\text{region } 1 + \text{region } 3) + (\text{region } 2 + \text{region } 3) - \text{region } 3$$

$$= \text{region } 1 + \text{region } 3 + \text{region } 2$$

$$= \text{region } 1 + \text{region } 2 + \text{region } 3,$$

it follows that

$$P(A \cup B) = P(A) + P(B) - P(A \cap B).$$

The additive rule can be extended to any number of events. For example, if event B is the union of events B_1 and B_2, then

$$P(B) = P(B_1 \cup B_2) = P(B_1) + P(B_2) - P(B_1 \cap B_2),$$

and

$$\begin{aligned} P(A \cup B) &= P(A) + P(B) - P(A \cap B) \\ &= P(A) + P(B_1) + P(B_2) - P(B_1 \cap B_2) \\ &\quad - P\big(A \cap (B_1 \cup B_2)\big). \end{aligned}$$

However,

$$P\big(A \cap (B_1 \cup B_2)\big) = P(A \cap B_1) + P(A \cap B_2) - P(A \cap B_1 \cap B_2).$$

(The last term is subtracted to prevent double counting.) Then

$$\begin{aligned} P(A \cup B_1 \cup B_2) &= P(A) + P(B_1) + P(B_2) - P(A \cap B_1) \\ &\quad - P(A \cap B_2) - P(B_1 \cap B_2) + P(A \cap B_1 \cap B_2). \end{aligned}$$

Mutually exclusive events

If $P(A \cap B) = 0$, then A and B cannot both occur, as illustrated in Figure 3. A and B are called *mutually exclusive events*.

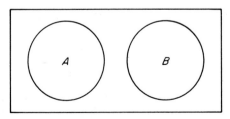

FIG. 3

If A and B are mutually exclusive events, the formula

$$P(A \cup B) = P(A) + P(B) - P(A \cap B)$$

reduces to

$$\boxed{P(A \cup B) = P(A) + P(B).}$$

Complementary events

For every event A there is an event $\bar{A}$, called the *complement of A*, with the two following properties:

(i) $P(A \cup \bar{A}) = 1$,
(ii) $P(A \cap \bar{A}) = 0$.

We already have the additive rule, $P(A \cup B) = P(A) + P(B) - P(A \cap B)$. If we let $B = \bar{A}$, then

$$P(A \cup \bar{A}) = P(A) + P(\bar{A}) - P(A \cap \bar{A})$$

which, using (i) and (ii) above, reduces to

$$\boxed{1 = P(A) + P(\bar{A}).}$$

This is shown in Figure 4, where region 1 is event A and region 2 is event $\bar{A}$. Events A and $\bar{A}$ are mutually exclusive events.

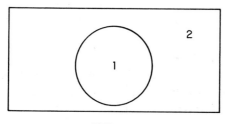

FIG. 4

If C is an event, then

$$P(C) = P(C \cap A) + P(C \cap \bar{A}),$$

as illustrated in Figure 5.

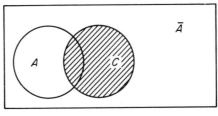

FIG. 5

Conditional probability

The probability that event B will occur given that event A has occurred is called a *conditional probability*, and is written as $P(B \mid A)$. In the Venn diagram in Figure 6, regions 1 and 3 represent event A, and regions

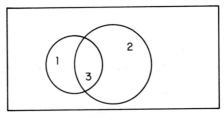

FIG. 6

2 and 3 event B. The probability of event B is the number of equally likely events in B divided by the total number of possible equally likely events. If we know that event A has occurred, the total number of possible events is the number of events in regions 1 and 3. Since A has oc-

curred, only those events B in regions 1 and 3 can occur. Then

$$P(B \mid A) = \frac{\text{number of events in 3}}{\text{number of events in 1 and 3}}$$

$$= \frac{\dfrac{\text{number of events in 3}}{\text{total number of possible events}}}{\dfrac{\text{number of events in 1 and 3}}{\text{total number of possible events}}} = \frac{P(A \cap B)}{P(A)}.$$

Hence, if $P(A)$ does not equal zero,

$$P(B \mid A) = \frac{P(A \cap B)}{P(A)} \quad \text{or} \quad P(B \mid A) \cdot P(A) = P(A \cap B).$$

One special case of dependence is called independence. Event B is said to be independent of event A if the probability of event B does not depend on whether event A has occurred. In symbolic notation, if $P(B) = P(B \mid A)$, then event B is independent of event A. If event B is independent of event A, then

$$P(A \cap B) = P(B \mid A) \cdot P(A) = P(B) \cdot P(A)$$

and

$$P(A \cap B) = P(A \mid B) \cdot P(B) = P(B) \cdot P(A).$$

If event B is independent of A, then A is independent of B. Thus, events A and B are *independent* if

$$P(A) \cdot P(B) = P(A \cap B),$$

There are two other types of dependence between two items. Events A and B are *complements* if event B is more likely to occur if event A has taken place. That is, A and B are complements if $P(B \mid A) > P(B)$.

Since, $P(B \mid A) \cdot P(A) = P(A \cap B)$, A and B are complements if

$$P(B) \cdot P(A) < P(A \cap B).$$

Events A and B are *substitutes* if event B is less likely to occur if event A has taken place, that is, if $P(B \mid A) < P(B)$. Since, $P(B \mid A) \cdot P(A) = P(A \cap B)$, events A and B are substitutes if

$$P(B) \cdot P(A) > P(A \cap B).$$

If events A and B are mutually exclusive events, we know that $P(A \cap B) = 0$. If both A and B have a nonzero probability of occurring, $P(A) \neq 0$ and $P(B) \neq 0$, and $P(A) \cdot P(B) > P(A \cap B)$. Thus mutually exclusive events are *not* independent events; they are substitute events.

Random Variables (Problems 27 through 35)

A *random variable* is a variable whose value is a number determined by the outcome of an experiment. For example, if we ask a group of individuals their incomes, the variable income is a random variable. A random variable has a *frequency distribution*. A frequency distribution indicates the number of observations (frequency, f_i) of a random variable (X) at each value of the random variable (X_i). Statements can be made about the probability of obtaining specific values of a variable (X) from random sampling if we know the characteristics of the variable's frequency distribution. Several frequency distributions are presented in this chapter.

Hypergeometric distribution

Suppose we have a population of N objects, some of which have a particular characteristic, N', and the others have a different characteristic, N'', where $N = N' + N''$. That is, we have a dichotomous population. For example, N may be people, N' males and N'' females. Or, N may be households, N' those households with a color TV and N'' those without a color TV. From the population (N) we choose a sample of size s. We want to know the probability that X observations in the sample of size s are of type N'. For example, what is the probability that there will be exactly 4 males (or 4 households with TV sets) in a sample of size 10 drawn from a population of size $N = N' + N''$? Once we sample an observation, we do not wish to select that observation again for the same problem; we do not put a sampled observation back into the pool from which we select observations.

Using the theory of combinations to be developed in the problems below, we find that

$$
P_H(X) = \frac{(N'C_x)(N''C_{s-x})}{NC_s} \quad \text{or} \quad \frac{\binom{N'}{x}\binom{N''}{s-x}}{\binom{N}{s}}.
$$

where $N'C_x$ or $\binom{N'}{x}$ is read as the number of different combinations of N' objects when samples of size x are selected.

The distribution of these probabilities for different values of X is called the hypergeometric distribution.

Binomial distribution

Suppose a sample of s observations is randomly selected with replacement of the observations from a dichotomous population in which some members have characteristic A and the others do not have this characteristic. Since each observation is replaced after it is sampled, the probability of selecting an observation with characteristic A does not change as more observations are sampled. Since P is the proportion of those observations in the population with characteristic A, P can be designated the probability of a "success." Since the population is dichotomous, $Q = 1 - P$ is the probability of a "failure," the probability that an observation does not have characteristic A.

The probability that a sample of size A will have x successes on the first x tries and $s - x$ failures on the next $s - x$ observations is $(P)^x(Q)^{s-x}$. However, if we are concerned only with x successes from a sample of size s without concern for the order of the successes, the probability is

$$P_B(x) = sC_x(P)^x(Q)^{s-x} = \binom{s}{x} P^x Q^{s-x}.$$

$P_B(x)$ is called the binomial probability, and the distribution of these probabilities for various values of x ($x = 0, 1, 2, \ldots, s$) is called the binomial distribution.

The hypergeometric and binomial distributions answer the same question: What is the probability of x "successes" (an observation with characteristic A) and $s - x$ "failures" if we sample from a dichotomous population of size N in which $N' = PN$ have characteristic A and $N'' = QN = (1 - P)N$ do not have this characteristic? However, these distributions differ in one respect. Under the binomial distribution the probability of a success is the same whether the 1st, 2nd, 3rd, or sth observation is selected because the observation is returned to the pool (replacement) after it is sampled. Under the hypergeometric distribution once an observation is sampled it is *not* returned (without replacement), so the composition of the population, and therefore the probability of a success, changes with successive sampling. If we are selecting a small sample from a very large population, there is a very small change in the probability of a success with each successive observation. Thus, if the sample is small relative to the population, the binomial and hypergeometric distributions give approximately the same probability distribution. Since it is easier to compute probabilities from the binomial than the hypergeometric distribution, this approximation is used when samples that are small relative to the population are selected without replacement. The table of values for the binomial distribution is Table 8 of the appendix.

Normal distribution

The normal distribution, sometimes called the bell-shaped or Gaussian distribution, is the most commonly used distribution in statistical analysis. The distribution is symmetric around the mean and has a known frequency distribution. A wide variety of frequency distributions observed in nature are close approximations to a normal distribution. The probability distribution obtained from the binomial distribution when $P = Q = 0.5$ is also normally distributed. Perhaps the most important theorem for statistical theory is the *Central Limit Theorem*. This theorem says that regardless of the shape of the distribution of independent random variables, a linear combination of these variables approaches the normal distribution as the number of variables in the linear combination increases. Since the arithmetic mean is a linear combination, the distribution of the sample means of a variable obtained from random sampling approaches the normal distribution for large samples.

If a variable X_i is normally distributed with a population mean μ and a population standard deviation σ, the variable Z_i, where $Z_i = (X_i - \mu)/\sigma$, has a *standardized normal distribution*. Z_i has a zero population mean and a unit standard deviation. The frequency distribution for the standardized normal distribution is

$$f(Z) = \frac{1}{\sqrt{2\pi}}\, e^{-Z_i{}^2/2},$$

where $\pi \approx \frac{22}{7} \approx 3.14$ and e is the base of the natural logarithm ($e \approx 2.718$) (Figure 7). Probability statements for the normal distribution are written as follows:

1. *One-tailed statement*:

$$P\left(\frac{X - \mu}{\sigma} > h\right) = a,$$

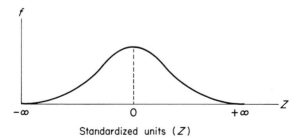

Standardized units (Z)

FIG. 7

which is read, "the probability that the value X exceeds the population mean (μ) by h standard deviations ($h\sigma$) is a."

2. *Two-tailed statement*:

$$P\left(-h < \frac{X - \mu}{\sigma} < +h\right) = b,$$

read, "the probability that the value X is closer to the population mean (μ) than h standard deviations ($h\sigma$) is b." (Since the normal distribution is symmetric around the mean, $b = 1 - 2a$.)

These two statements are shown graphically in Figure 8.

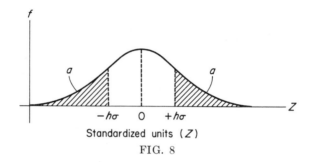

Standardized units (Z)

FIG. 8

The probability that a randomly selected observation from a normal distribution will be more than h standard deviations away from the population mean is:

h	Probability†
1	0.3174
2	0.0456
3	0.0026

For example, if height is normally distributed only 2.28 percent of the population are two standard deviations or more taller than the mean

---

† See Table I in the Appendix for the computation of probabilities for a normal distribution.

and 2.28 percent are two standard deviations or more shorter than the mean.

Chebyshev's inequality

There are situations in which the shape of the frequency distribution under consideration is not known. However, this does not mean that no probability statements can be made concerning how far above the mean a randomly selected observation will be located.

Chebyshev's Theorem‡ states that

$$
P\left(\left|\frac{X - \mu}{\sigma}\right| \geq h\right) \leq \frac{1}{h^2}.
$$

That is, the probability that a random variable (X) from a distribution of unknown shape differs from its mean (μ) by h standard deviations (σ) or more is less than or equal to $1/h^2$:

h	Probability is less than or equal to
1	1.0000
2	0.2500
3	0.1111

Note that 4.56 percent of the observations are two standard deviations or more away from the mean if the distribution is known to be normally distributed, but 25 percent or less are two standard deviations or more from the mean if the variable's distribution is unknown. Thus the knowledge that a variable is normally distributed is a very valuable piece of information. The Central Limit Theorem, which tells us that means of observations from large samples are approximately normally distributed, is an important theorem in statistical analysis.

‡ For a proof of the theorem see F. Mosteller, R. E. K. Rourke, and G. B. Thomas, *Probability With Statistical Applications* (Addison-Wesley, Reading, Mass., 1961), pp. 203–205.

Problems

1. Number of Arrangements—Tree Diagram
2. Number of Arrangements
3. Number of Arrangements
4. Number of Arrangements
5. Factorials
6. Permutations
7. Combinations
8. Combinations
9. Perturbations and Combinations
10. Combinations
11. Perturbations
12. Permutations, N_1 Identical Objects
13. Permutations, N_1 Identical and N_2 Identical Objects
14. Permutations
15. Probability
16. Probability
17. Probability
18. Probability—Additive Rule
19. Mutually Exclusive Events
20. Sums of Mutually Exclusive Events
21. Intersection of Events, without Replacement
22. Intersection of Events, with Replacement
23. Complementary Events
24. Independent and Dependent Events
25. Conditional Probability
26. Conditional Probability
27. Hypergeometric and Binomial Probability Distributions
28. Chebyshev's Inequality
29. Chebyshev's Inequality
30. Normal Distribution
31. Normal Distribution
32. Normal Distribution
33. Binomial and Normal Distributions
34. Poisson Distribution
35. Geometric Distribution

Problem 1. Number of Arrangements—Tree Diagram

Q: Mary and John enter a room with five seats. In how many ways may they be seated?

A: A fundamental principle of counting states that if (i) event A may occur in r ways and (ii) for *each* way in which event A may occur, event B may occur in s ways, *then* events A and B may *both* occur in rs ways.

    John has five choices for a seat. (Event A may occur in five ways.) Whatever seat John chooses, Mary has four choices for a seat. (For each way in which event A may occur, event B may occur in four ways.) Therefore $5 \cdot 4 = 20$ is the number of ways in which John and Mary may be seated. A tree diagram of the 20 possible seating arrangements is given in Figure 9.

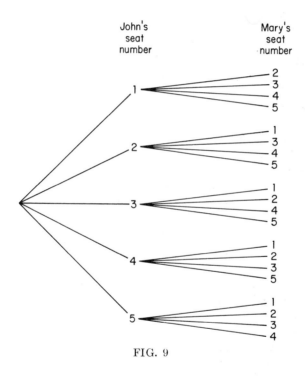

FIG. 9

Problem 2. Number of Arrangements

Q: In how many different ways can four workers be allocated among nine factories if no two workers go to the same factory?

A: The first worker can go to any one of nine factories (event A). After he is assigned, the second worker can go to any one of eight (event B); the third worker any one of seven (event C); and last, the fourth worker any one of six (event D). Since

$9 \cdot 8 \cdot 7 \cdot 6 = 3{,}024$,

there are 3,024 ways for four workers to be allocated among nine factories if no two workers go to the same factory.

Problem 3. Number of Arrangements

Q: A firm wishes to identify its products by four-digit codes, where the first digit is a letter other than I or O, and the remaining three digits are numbers, but the numbers cannot all be zeros. How many different products can be identified this way?

A: For the first digit there are $26 - 2 = 24$ possible choices. For the second, third, and fourth digits there are $10 \cdot 10 \cdot 10 - 1$ possibilities, 10 digits in each of the three slots (assuming digits can be repeated) minus one possibility for three zeros. Therefore there are $24 \cdot 99 = 23{,}976$ classification numbers available.

Problem 4. Number of Arrangements

Q: Bob has five brothers. In how many ways can Bob invite one or more of his brothers to his home?

A: For each brother there are two choices: "invited" or "not invited." For two brothers, John and Tom, there are four choices:

John	Tom
invited	invited
invited	not invited
not invited	invited
not invited	not invited

For all five brothers there are $2 \cdot 2 \cdot 2 \cdot 2 \cdot 2 = 2^5 = 32$ possibilities. However, among the 32 possibilities is the one in which none of the brothers is invited. Since Bob is inviting *one or more* brothers to his home we must subtract 1 from 32. Bob may accomplish his task in 31 different ways.

Problem 5. Factorials

Q: If N is a positive integer (whole number), N factorial ($N!$) is the product of the first N positive integers.
(a) Write the expression for N factorial ($N!$).
(b) What is $4!$? $6!$? $1!$? $0!$?

A: (a) If $N!$ (N factorial) is the product of the first N positive integers, then

$$N! = N \cdot (N - 1) \cdot (N - 2) \cdots 4 \cdot 3 \cdot 2 \cdot 1.$$

(b) Using the formula, we get

$4! = 4 \cdot 3 \cdot 2 \cdot 1 = 24,$
$6! = 6 \cdot 5 \cdot 4 \cdot 3 \cdot 2 \cdot 1 = 720,$
$1! = 1.$

By definition $0! = 1$.
Appendix Table 10 presents the solutions for many factorials.

Problem 6. Permutations

Q: Two positions are empty on a book shelf. In how many ways can the positions be filled if there are five books available? Assume that the books are different colors (red, green, blue, yellow, and orange) and that the order of the books on the shelf matters.

A: This is an example of a problem involving *permutations*. We are interested in the number of permutations of five items taken two at a time.
The notation is $5P_2$ or $P_2{}^5$.

$$5P_2 = \frac{5!}{(5 - 2)!} = \frac{5!}{3!} = \frac{5 \cdot 4 \cdot 3 \cdot 2 \cdot 1}{3 \cdot 2 \cdot 1} = 5 \cdot 4 = 20.$$

A tree diagram of the 20 permutations of the five books taken two at a time is shown in Figure 10. Note that for permutations order is important. For example, red–orange (4) is different from orange–red (17).

Slot 1   Slot 2

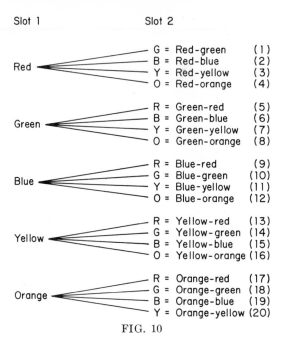

FIG. 10

Problem 7. Combinations

Q: Solve the preceding problem but assume that the order of the two books on the shelf does *not* matter.

A: If order is unimportant, we are interested in the number of *combinations* of five books taken two at a time.

The number of combinations of five things taken two at a time is written as $5C_2$ or C_2^5 or $\binom{5}{2}$.

$$5C_2 = \frac{5!}{2!(5-2)!} = \frac{5!}{2!\,3!} = \frac{5 \cdot 4 \cdot 3 \cdot 2 \cdot 1}{2 \cdot 1 \cdot 3 \cdot 2 \cdot 1} = \frac{5 \cdot 4}{2 \cdot 1} = 10.$$

The ten combinations of the five books taken two at a time are:

1. red–green or green–red,
2. red–blue or blue–red,
3. red–yellow or yellow–red,
4. red–orange or orange–red,

5. blue–green or green–blue,
6. blue–yellow or yellow–blue,
7. blue–orange or orange–blue,
8. green–yellow or yellow–green,
9. green–orange or orange–green,
10. yellow–orange or orange–yellow.

Problem 8. Combinations

Q: A club has seven members. How many committees of three members each can be formed from the club?

A: Since there is no reason to order the three members of the committee, we are interested in the number of *combinations* of seven people taken three at a time.

$$7C_3 = \frac{7!}{3!(7-3)!} = \frac{7!}{3!\,4!} = \frac{7 \cdot 6 \cdot 5 \cdot 4 \cdot 3 \cdot 2 \cdot 1}{3 \cdot 2 \cdot 1 \cdot 4 \cdot 3 \cdot 2 \cdot 1} = \frac{7 \cdot 6 \cdot 5}{3 \cdot 2 \cdot 1} = 35.$$

The number of committees that can be formed is 35.

Problem 9. Permutations

Q: Do the previous problem with the additional information that each committee of three is to have a president, treasurer, and secretary. We now have reason to order the committees, and we are interested in the number of *permutations* of seven people taken three at a time.

A: The number of permutations for this problem is $7P_3$:

$$7P_3 = \frac{7!}{(7-3)!} = \frac{7!}{4!} = \frac{7 \cdot 6 \cdot 5 \cdot 4 \cdot 3 \cdot 2 \cdot 1}{4 \cdot 3 \cdot 2 \cdot 1} = 7 \cdot 6 \cdot 5 = 210.$$

Note: Notice the difference in the formulas for Problems 8 and 9. There are many more permutations (with ordering) than combinations (without ordering). Three members of the club (Bob, John, and Tom) form *one* committee. However, in how many ways may we choose a president, treasurer, and secretary from the three men?

There are *six* ways of accomplishing this task:

President	Treasurer	Secretary
Bob	John	Tom
Bob	Tom	John
John	Bob	Tom
John	Tom	Bob
Tom	John	Bob
Tom	Bob	John

$$7P_3 = \frac{7!}{(7-3)!} \quad \text{and} \quad 7C_3 = \frac{7!}{3!(7-3)!}.$$

The formulas differ by a factor of $3! = 6$ in the denominator of the representation of $7C_3$.

Problem 10. Combinations

Q: Three cards are selected from a deck of 52 playing cards. What is the probability that

(a) all three are aces?
(b) at least one is an ace?
(c) the three cards come from different suits?

A: If it is assumed that order is not important and that the cards are drawn without replacement (i.e., a card is not returned to the deck after it is drawn), then the total number of possible outcomes is

$$52C_3 = \frac{52!}{3!(52-3)!} = \frac{52 \cdot 51 \cdot 50}{3 \cdot 2 \cdot 1} = 22,100.$$

The probability of a success is the number of successes divided by the total number of possible outcomes.

(a) The number of outcomes in which all three cards are aces is

$$4C_3 = \frac{4!}{3!\,1!} = 4$$

since there are only four aces. The probability of obtaining three aces is

$$\frac{4C_3}{52C_3} = \frac{4 \cdot 3 \cdot 2}{52 \cdot 51 \cdot 50} = 0.00018.$$

(b) The probability of drawing at least one ace equals one minus the probability of drawing no aces: P(at least one ace) = $1 - P$(no aces). The number of situations of no aces is $48C_3$ since there are 48 non-aces:

$$48C_3 = \frac{48!}{3! \, 45!} = \frac{48 \cdot 47 \cdot 46}{1 \cdot 2 \cdot 3} = 17{,}296.$$

The probability of no aces is $48C_3/52C_3$. The probability of drawing at least one ace is therefore

$$1 - \frac{48C_3}{52C_3} = 1 - \frac{48 \cdot 47 \cdot 46}{52 \cdot 51 \cdot 50} = 1 - 0.7826 = 0.2174.$$

(c) There are

$$13C_1 = \frac{13!}{1! \, 12!} = 13$$

ways of getting a card from a particular suit on a given draw. There are $4C_3 \cdot 13C_1 \cdot 13C_1 \cdot 13C_1 = 4 \cdot 13 \cdot 13 \cdot 13 = 8788$ ways of getting three cards each of different suits from the four suits. The probability of this outcome is

$$\frac{4C_3 \cdot 13C_1 \cdot 13C_1 \cdot 13C_1}{52C_3} = 0.3976.$$

Note: If there is *replacement* of each card after it is drawn, the answers are

(a) $(\frac{4}{52})^3 = 0.00046,$
(b) $1 - (\frac{48}{52})^3 = 0.2135,$ and

(c) there are 4! ways of organizing the three cards, and the probability of drawing a card from a particular suit is $\frac{1}{4}$.

$$4! \cdot \frac{1}{4} \cdot \frac{1}{4} \cdot \frac{1}{4} = 0.375.$$

Does "replacement" make much of a difference?

Problem 11. Permutations

Q: How many arrangements are there of the letters of the word "bark"?

A: The word "bark" has four different letters: a, b, k, r. To form a "word" from these letters we have four choices for the first position. For each of these choices we have three choices for the second position. For each choice of first and second position we have two choices for the third position. For each choice of first, second, and third position we have one choice for the fourth position.

We are interested in the number of permutations of four items taken four at a time. This is $4P_4$.

$$4P_4 = \frac{4!}{(4-4)!} = \frac{4!}{0!} = \frac{4!}{1} = 4! = 4 \cdot 3 \cdot 2 \cdot 1 = 24.$$

There are 24 arrangements of the letters of the word "bark." They are:

1. $a\,b\,k\,r$	9. $b\,k\,a\,r$	17. $k\,r\,a\,b$
2. $a\,b\,r\,k$	10. $b\,k\,r\,a$	18. $k\,r\,b\,a$
3. $a\,k\,b\,r$	11. $b\,r\,a\,k$	19. $r\,a\,b\,k$
4. $a\,k\,r\,b$	12. $b\,r\,k\,a$	20. $r\,a\,k\,b$
5. $a\,r\,b\,k$	13. $k\,a\,b\,r$	21. $r\,b\,a\,k$
6. $a\,r\,k\,b$	14. $k\,a\,r\,b$	22. $r\,b\,k\,a$
7. $b\,a\,k\,r$	15. $k\,b\,a\,r$	23. $r\,k\,a\,b$
8. $b\,a\,r\,k$	16. $k\,b\,r\,a$	24. $r\,k\,b\,a$

Problem 12. Permutations, N_1 Identical Objects

Q: How many arrangements are there of the letters of the word "book"?

A: *If* we distinguished between the two o's, calling them o_1 and o_2, we would have $4! = 24$ arrangements. (See the previous problem.)

We are, however, *not* going to distinguish between the two o's. o_1 and o_2 may be arranged in 2! = 2 ways: o_1o_2 and o_2o_1. If we divide 4! by 2! we obtain the desired answer. This problem involves the number of permutations of four items of which two are identical.

$$\frac{4!}{2!} = \frac{4\cdot3\cdot2\cdot1}{2\cdot1} = 12.$$

The 12 arrangements of the letters of the word "book" are:

1. *b k o o*	5. *o b k o*	9. *o o b k*
2. *b o k o*	6. *o b o k*	10. *k o o b*
3. *b o o k*	7. *k o b o*	11. *o k o b*
4. *k b o o*	8. *o k b o*	12. *o o k b*

Problem 13. Permutations, N_1 Identical and N_2 Identical Objects

Q: How many arrangements are there of the letters of the word "missions"?

A: The word "missions" has eight letters. The two i's may be arranged in 2! = 2 ways: i_1i_2 and i_2i_1. The three s's may be arranged in 3! = 6 ways: $s_1s_2s_3$, $s_1s_3s_2$, $s_2s_1s_3$, $s_2s_3s_1$, $s_3s_1s_2$, and $s_3s_2s_1$. Dividing, we get

$$\frac{8!}{2!\,3!} = \frac{8\cdot7\cdot6\cdot5\cdot4\cdot3\cdot2\cdot1}{2\cdot1\cdot3\cdot2\cdot1} = 3{,}360.$$

There are 3,360 arrangements of the letters of the word "missions."

Note: If we wish to arrange in order N items of which N_1 are indistinguishable from each other and N_2 other items are indistinguishable from each other ($N_1 + N_2$ is less than or equal to N), we can have

$$\frac{N!}{N_1!\,N_2!}$$

different arrangements.

Problem 14. Permutations

Q: The Smith family consists of eight members; the Johnson family consists of ten members. In how many different ways can an interviewer arrange separate telephone interviews with six individuals if three are from the Smith family and three are from the Johnson family? (Assume that the order of individuals matters.)

A: There are $8P_3 = 8 \cdot 7 \cdot 6 = 336$ ways of ordering the interviewing of the Smiths, and $10P_3 = 10 \cdot 9 \cdot 8 = 720$ ways of interviewing the Johnsons. The number of different ways of having two separate samples of three Smiths and three Johnsons is $366 \cdot 720$. These two samples can be ordered in $(6!/3!\,3!) = 20$ different ways. (That is, the number of permutations of n objects of which n_i are of one kind and n_j are of another kind is $n!/n_i!\,n_j!$.) The answer is

$$366 \cdot 720 \cdot 20 = 5,270,400.$$

Problem 15. Probability

Q: An urn contains only two red balls. One ball is drawn from the urn. What is the probability that this ball is (a) red? (b) blue?

A: (a) $P(\text{ball is red}) = \dfrac{\text{number of red balls}}{\text{total number of balls}} = \dfrac{2}{2} = 1.$

(b) $P(\text{ball is blue}) = \dfrac{\text{number of blue balls}}{\text{total number of balls}} = \dfrac{0}{2} = 0.$

The probability of an event A is the ratio of the number of equally likely possible outcomes of event A to the total number of possible outcomes.

Problem 16. Probability

Q: An urn contains two red and three blue balls. One ball is drawn from the urn. What is the probability that it is (a) blue? (b) red? (c) either red or blue?

A: (a) $P(\text{blue}) = \dfrac{\text{number of blue balls}}{\text{total number of balls}},$

$P(\text{blue}) = \frac{3}{5} = 0.6.$

(b) $P(\text{red}) = \dfrac{\text{number of red balls}}{\text{total number of balls}}$,

$P(\text{red}) = \frac{2}{5} = 0.4$.

(c) $P(\text{red or blue}) = \dfrac{\text{number of red or blue balls}}{\text{total number of balls}}$,

$P(\text{red or blue}) = \frac{5}{5} = 1.0$.

Problem 17. Probability

Q: One card is drawn from a deck of 52 playing cards. What is the probability that it is (a) a three? (b) a diamond? (c) a three of diamonds?

A: (a) $P(\text{three}) = \dfrac{\text{number of threes}}{\text{total number of cards}}$,

$P(\text{three}) = \frac{4}{52} = \frac{1}{13}$.

(b) $P(\text{diamond}) = \dfrac{\text{number of diamonds}}{\text{total number of cards}}$,

$P(\text{diamond}) = \frac{13}{52} = \frac{1}{4}$.

(c) $P(\text{three of diamonds}) = \dfrac{\text{number of "three of diamonds"}}{\text{total number of cards}}$,

$P(\text{three of diamonds}) = \frac{1}{52}$.

Problem 18. Probability—Additive Rule

Q: One card is drawn from a deck of 52 playing cards. What is the probability that the card is (a) a three and a diamond? (b) a three or a diamond or both?

A: Only one card in the deck is a three and a diamond: the three of diamonds. Therefore $P(\text{three and diamond}) = \frac{1}{52}$.

There are four threes and thirteen diamonds in a deck: $4 + 13 = 17$. We must remember, however, that the three of diamonds has been counted twice, as both a three and a diamond: $17 - 1 = 16$. Therefore, 16 cards are either a three or a diamond or both:

$$P(\text{three or diamond or both}) = \tfrac{16}{52} = \tfrac{4}{13}.$$

Note:

$P(\text{three or diamond or both}) = P(\text{three} \cup \text{diamond}),$
$P(\text{three and diamond}) = P(\text{three} \cap \text{diamond}).$

From the answers to the two previous problems we have the information:

(i) $P(\text{diamond}) = \tfrac{13}{52},$
(ii) $P(\text{three}) = \tfrac{4}{52},$
(iii) $P(\text{three} \cap \text{diamond}) = \tfrac{1}{52},$
(iv) $P(\text{three} \cup \text{diamond}) = \tfrac{16}{52}.$

Thus the formula

$$P(A \cup B) = P(A) + P(B) - P(A \cap B)$$

is verified:

$$P(\text{three} \cup \text{diamond}) = P(\text{three}) + P(\text{diamond})$$
$$- P(\text{three} \cap \text{diamond}),$$

or, substituting,

$$\tfrac{16}{52} = \tfrac{4}{52} + \tfrac{13}{52} - \tfrac{1}{52} = \tfrac{16}{52} = \tfrac{4}{13}.$$

Problem 19. Mutually Exclusive Events

Q: One card is drawn from a deck of 52 playing cards. What is the probability that the card is a three or a four?

A: The probability that the card is both a three and a four is zero: $P(\text{three} \cap \text{four}) = 0$. Therefore, picking a three and picking a four are mutually exclusive events:

$$P(\text{three} \cup \text{four}) = P(\text{three}) + P(\text{four}),$$

$$P(\text{three} \cup \text{four}) = \tfrac{4}{52} + \tfrac{4}{52} = \tfrac{8}{52} = \tfrac{2}{13}.$$

Problem 20. Sums of Mutually Exclusive Events

Q: A committee of three—A, B, and C—is to make a decision on the basis of a majority vote. What is the probability of a wrong decision by the committee if the probabilities of a wrong decision by each member are $P(A) = 0.05$, $P(B) = 0.05$, and $P(C) = 0.10$?

A: There are eight possible outcomes of a vote (if order is ignored), but only four result in a wrong decision. (Use a tree diagram to show this.)

$$P(A, B, \text{ and } C \text{ wrong}) = 0.05 \cdot 0.05 \cdot 0.10 = 0.00025,$$

if each votes independently.
Also

$$P(A \text{ and } B \text{ wrong}, C \text{ right}) = 0.05 \cdot 0.05 \cdot 0.9 = 0.00225,$$

$$P(A \text{ and } C \text{ wrong}, B \text{ right}) = 0.05 \cdot 0.10 \cdot 0.95 = 0.00475,$$

$$P(B \text{ and } C \text{ wrong}, A \text{ right}) = 0.05 \cdot 0.10 \cdot 0.95 = 0.00475.$$

The four outcomes are mutually exclusive. Therefore the probability of a wrong decision by the committee is the sum of the four probabilities or 0.01200. The committee will be wrong in 1.2 percent of its decisions.

Problem 21. Intersection of Events, without Replacement

Q: An urn contains two red and three green balls. Let event A be drawing a red ball from the urn on the first draw. Let event B be drawing a green ball from the urn on the second draw. Find the probability that both event A and event B occur: $P(A \cap B)$.

A: Using the formula relating the intersection of two events to the conditional probability, $P(A \cap B) = P(B \mid A)P(A)$.

$$P(A) = \tfrac{2}{5} \quad \text{and} \quad P(B \mid A) = \tfrac{3}{4}.$$

(On the second draw the urn contains only four balls.) Therefore

$$P(A \cap B) = \left(\tfrac{3}{4}\right)\left(\tfrac{2}{5}\right) = \tfrac{3}{10}.$$

The probability is 0.3 that both events occur.

Problem 22. Intersection of Events, with Replacement

Q: An urn contains two red and three green balls. Let event A be drawing a red ball from the urn on the first draw. The red ball is immediately returned to the urn. Let event B be drawing a green ball from the urn on the second draw. Find $P(A \cap B)$.

A: Since the red ball picked on the first draw is immediately returned to the urn, A and B are independent events. Therefore

$$P(A \cap B) = P(A)\, P(B).$$

Then

$$P(A) = \tfrac{2}{5}, \qquad P(B) = \tfrac{3}{5},$$

and

$$P(A)\, P(B) = \tfrac{2}{5} \cdot \tfrac{3}{5} = \tfrac{6}{25},$$
$$P(A \cap B) = \tfrac{6}{25}.$$

The probability is 0.24 that events A and B will both occur.

Problem 23. Complementary Events

Q: The probability of event C is $\tfrac{2}{3}$. What is the probability that event C does *not* occur?

A: $P(C) + P(\bar{C}) = 1,$

$\tfrac{2}{3} + P(\bar{C}) = 1,$

$P(\bar{C}) = 1 - \tfrac{2}{3} = \tfrac{1}{3}.$

Problem 24. Independent and Dependent Events

Q: The relative frequency distribution of work status and presence of children in the household of a random sample of 1,000 married women is:

	Works	Doesn't work	Total
Children	0.1	0.4	0.5
No children	0.3	0.2	0.5
Total	0.4	0.6	1.0

(a) Are children and work mutually exclusive?
(b) Is work related to the presence of children?
(c) Are work and children substitutes or complements?

A: (a) No. If children and work were mutually exclusive there would be no observations in the "children—works" cell.

(b) If two events A and B are independent, $P(A) = P(A \mid B)$. We know that $P(A \mid B) = P(A \cap B)/P(B)$. If A = works and B = children, $P(A \cap B) = 0.1$ (i.e., the probability that both A and B occur), and $P(B) = 0.5$. Therefore $P(A \mid B) = 0.2$. However, $P(A) = 0.4$ and $P(A) \neq P(A \mid B)$. A and B are not independent. A woman's working and the presence of children are related.

(c) $P(A \cap B) = P(A \mid B) P(B) = P(B \mid A) P(A)$. For independence, $P(A \cap B) = P(A)P(B)$. For substitutes $P(A \cap B) < P(A)P(B)$. (The product of the separate probabilities is greater than the joint probability.) For complements $P(A \cap B) > P(A)P(B)$. $P(A \cap B) = 0.1$ and $P(A)P(B) = 0.2$. Therefore work and children are substitutes. A woman with children has a lower probability of working than a woman without children.

Problem 25. Conditional Probability

Q: The output of three factories A, B, and C all goes to the same warehouse. The proportion of total output from each factory and the proportion of defective parts in that factory's output are:

	A	B	C
Output (%)	0.30	0.25	0.45
Defective parts (%)	0.01	0.12	0.02

What is the probability that a randomly selected item is defective?

A: The probability of a defective part $[P(D)]$ is the probability that the part is defective and from factory A $[P(A \cap D)]$, plus the probability that it is defective and from factory B $[P(B \cap D)]$, plus the probability that it is defective and from factory C $[P(C \cap D)]$:

$$
\begin{aligned}
P(D) &= P(A \cap D) + P(B \cap D) + P(C \cap D) \\
&= P(D \mid A)P(A) + P(D \mid B)P(B) + P(D \mid C)P(C) \\
&= 0.01 \cdot 0.3 + 0.12 \cdot 0.25 + 0.02 \cdot 0.45 \\
&= 0.042.
\end{aligned}
$$

The probability is 0.042 that a randomly selected part is defective; or, 4.2 percent of the parts are defective.

Problem 26. Conditional Probability

Q: A medical test is negative for 95 percent of the population. It is known that 90 percent of those for whom the test is positive and 20 percent of those for whom the test is negative do in fact have the disease.

    (a) What is the probability that a randomly tested person has the disease?

    (b) If a person has the disease, what is the probability that he would react negatively to the test?

A: (a) Let H designate a negative result to the test, $\bar{H}$ a positive result, and B having the disease. Then

$$P(B) = P(B \cap H) + P(B \cap \bar{H})$$
$$= P(B \mid H)P(H) + P(B \mid \bar{H})P(\bar{H})$$
$$= 0.20 \cdot 0.95 + 0.90 \cdot 0.05 = 0.235.$$

(b) If a person has the disease, the probability that his test result would be negative is

$$P(H \mid B) = \frac{P(H \cap B)}{P(B)} = \frac{0.20 \cdot 0.95}{0.235} = 0.81,$$

That is, 81 percent of those who have the disease would have negative test results.

Problem 27. Hypergeometric and Binomial Probability Distributions

Q: What is the probability of obtaining two men and four women by drawing, without replacement, six people at random from

(a) a group of five men and five women?
(b) a group of 500 men and 500 women?

A: (a) For population values $M = 5$, $W = 5$, and $N = M + W = 10$. For the sample, $m = 2$, $w = 4$, and $n = 6$.

$$P_H = \frac{\left(\begin{array}{l}\text{Number of ways of} \\ \text{having a sample of 2} \\ \text{from a population of 5}\end{array}\right)\left(\begin{array}{l}\text{Number of ways of} \\ \text{having a sample of 4} \\ \text{from a population of 5}\end{array}\right)}{\left(\begin{array}{l}\text{Number of ways of having a sample} \\ \text{of 6 from a population of 10}\end{array}\right)} = \frac{\binom{5}{2}\binom{5}{4}}{\binom{10}{6}}$$

$$= \frac{5}{21} = 0.2381.$$

(The subscript H denotes hypergeometric distribution.)

(b) $P_H = \dfrac{\dbinom{500}{2}\dbinom{500}{4}}{\dbinom{1000}{6}} \approx 0.2346.$

Note: If the binomial distribution were used (i.e., if the sampling were done with replacement), then

$$P_B = \binom{n}{m}(P_m)^m (P_w)^w$$

$$= \binom{6}{2}\left(\frac{1}{2}\right)^2 \left(\frac{1}{2}\right)^4 \approx 0.2344.$$

However,

$$P_H = \dfrac{\left(\dfrac{500!}{2!\,498!}\right)\left(\dfrac{500!}{4!\,496!}\right)}{\dfrac{1000!}{6!\,994!}} = \dfrac{6!}{4!\,2!}\left(\dfrac{\left(\dfrac{500!}{498!}\right)\left(\dfrac{500!}{496!}\right)}{\dfrac{1000!}{994!}}\right)$$

$$\approx 0.2346$$

and

$$P_B = \left(\frac{6!}{4!\,2!}\right)\left(\frac{(500)^2}{(1000)^2}\right)\left(\frac{(500)^4}{(1000)^4}\right) \approx 0.2344.$$

Since

$$\left(\frac{500!}{498!}\right)\left(\frac{500!}{496!}\right) \approx (500)^6 \quad \text{and} \quad \frac{1000!}{994!} \approx (1000)^6,$$

it follows that $P_B \approx P_H$. Thus when the population is large relative to the sample, sampling with replacement (binomial distribution) and sampling without replacement (hypergeometric distribution) give similar probabilities.

Problem 28. Chebyshev's Inequality

Q: If a student has a grade point average three standard deviations above the mean in his school, what proportion of his fellow students have higher grades?

A: The question does not specify the shape of the distribution of the students' grades. We can, however, apply Chebyshev's Theorem, which holds for any distribution.

Let X_i be the individual student's grade point average, μ the mean value, and σ the standard deviation. Chebyshev's Theorem states that

$$ P\left(\frac{|X_i - \mu|}{\sigma} \geq h\right) \leq \frac{1}{h^2} $$

or that the probability that a grade is more than h standard deviations away from the mean is less than or equal to $1/h^2$. Thus at most $1/h^2 = 1/3^2 = 0.11$, or 11 percent of his fellow students have higher averages.

Problem 29. Chebyshev's Inequality

Q: For a population the mean level of schooling is seven years and the standard deviation is one year. What is the probability that a randomly selected individual has between five and nine years of schooling?

A: Using Chebyshev's Theorem, we have

$$ P\left(\frac{|X_i - \mu|}{\sigma} \geq h\right) \leq \frac{1}{h^2} . $$

This is easily converted into a statement about the probability that a randomly selected observation from an unknown distribution is *within* h standard deviations of the mean. Since all observations must be within h standard deviations of the mean, or further away from the mean than h standard deviations,

$$\left(\frac{1}{h^2}\right) + \left(1 - \frac{1}{h^2}\right) = 1.$$

Then,

$$P\left(\frac{|X_i - \mu|}{\sigma} < h\right) \geq 1 - \left(\frac{1}{h}\right)^2,$$

$$P\left(\frac{|X_i - 7|}{1} < 2\right) \geq 1 - \left(\frac{1}{2}\right)^2 = 0.75.$$

The probability is at least 0.75 that the schooling level is within two standard deviations of the mean.

Problem 30. Normal Distribution

Q: Solve Problems 28 and 29 assuming the distribution in the population is normal. (Use Appendix Table 1 to find the area under the normal distribution.)

A: Using the normal distribution table, we find that

(a) $P\left(\frac{X_i - \mu}{\sigma} > 3\right) = 0.0013.$

(b) If $\mu = 7$ and $\sigma = 1$, then,

$$P(5 < X_i < 9) = P\left(\frac{5 - 7}{1} < \frac{X_i - \mu}{\sigma} < \frac{9 - 7}{1}\right)$$

$$= P\left(-2 < \frac{X_i - \mu}{\sigma} < +2\right) = 0.9544.$$

$(X_i - \mu)/\sigma$ is the standardized normal variate Z (see Figure 11). Note the change in the probability when the population

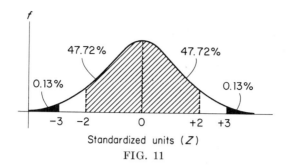

Standardized units (Z)

FIG. 11

is known to be normally distributed rather than to have an unknown distribution.

Problem 31. Normal Distribution

Q: Assume income is normally distributed with $\mu = \$5,000$ and $\sigma = \$1,000$. What proportion of the population has income between $\$4,000$ and $\$7,000$?

A: $P(\$4,000 < X < 7,000)$

$$= P\left(\frac{4,000 - 5,000}{1,000} < \frac{X - \mu}{\sigma} < \frac{7,000 - 5,000}{1,000}\right)$$

$$= P\left(-1 < \frac{X - \mu}{\sigma} < +2\right)$$

$$= P(-1 < X < 0) + P(0 < X < +2)$$

$$= 0.3413 + 0.4772 = 0.8185.$$

Therefore 81.85 percent of the population falls in the interval (see Figure 12).

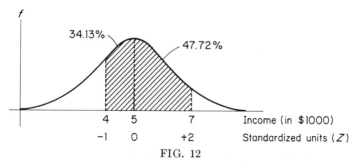

FIG. 12

Problem 32. Normal Distribution

Q: Over a large number of years the average daily number of passengers on a given airline is 10,000 and the standard deviation is 500. In one month (30 days) the average daily number of passengers was 9,400. What is the probability of deviating this far (or more) below the population mean?

A: A sample of size 30 is sufficiently large for the Central Limit Theorem to indicate that the sample mean is approximately normally distributed. If the question referred to only one week ($N = 7$), the normal distribution would not apply and Chebyshev's Inequality would be used.

Let Z be the standardized normal variate:

$$Z = \frac{\bar{X} - \mu}{\sigma_{\bar{x}}} = \frac{(\bar{X} - \mu)\sqrt{N}}{\sigma_x} = \frac{(9{,}400 - 10{,}000)\sqrt{30}}{500},$$

$$Z = -6.58.$$

Then

$$P\left(\frac{X - \mu}{\sigma_{\bar{x}}} \le -6.85\right) \approx 0.$$

The probability that an observation will be more than 6.58 standard deviations below the mean is approximately zero.

Note: If $\bar{X} = \dfrac{1}{n}\sum\limits_{i=1}^{n} X_i$ for a random sample, and the observations are statistically independent of each other,

$$\boxed{\sigma_{\bar{x}}^2 = \frac{1}{n^2}\sum_{i=1}^{n}\sigma_{x_i}^2 = \frac{n}{n^2}\sigma_{x_i}^2 = \frac{\sigma_x^2}{n}.}$$

Problem 33. Binomial and Normal Distributions

Q: Given that the probability of passing an exam is 0.75, what is the probability of

(a) passing at least five exams if you take ten?

(b) passing at least 12 exams if you take 24?

(See Appendix Table 8 for the values of the binomial distribution.)

A: (a) The probability of passing at least five exams is the sum of the probabilities of passing 5, 6, 7, 8, 9, and 10 exams.

The probability of passing only six exams in one particular order is $(0.75)^6(0.25)^4$. Since order doesn't matter and there are

$$\binom{10}{6} = \frac{10!}{6!\,4!}$$

different ways of passing six exams, the probability of passing six exams is

$$\binom{10}{6}(0.75)^6(0.25)^4.$$

The probability of passing m exams is

$$\binom{10}{m}(0.75)^m(0.25)^{10-m}.$$

The probability of passing at least five exams is therefore

$$P(\text{passing at least 5}) = \sum_{j=5}^{10}\binom{10}{j}(0.75)^j(0.25)^{10-j},$$

or, using the relation

$$P(\text{passing at least 5}) = 1.0 - P(\text{passing at most 4})$$

$$= 1.0 - \sum_{j=0}^{4}\binom{10}{j}(0.75)^j(0.25)^{10-j}$$

$P(\text{passing at least } 5) = 1.0 - 0.020 = 0.98.$

(b) $P(\text{passing } m \text{ exams}) = \binom{24}{m}(0.75)^m (0.25)^{24-m},$

$$P(\text{passing at most } 11) = \sum_{j=0}^{11} \binom{24}{j}(0.75)^j(0.25)^{24-j} = 0.002,$$

$P(\text{passing at least } 12) = 1.0 - 0.002 = 0.998.$

Note: A sample of 24 is almost large enough to enable us to use the normal distribution as an approximation to the binomial distribution. A sample size of at least 30 is the typical rule of thumb. Also, if N is large but the probability of success is very small or very large so that $NP < 10$ or $N(1 - P) < 10$, the normal approximation does not apply, but the Poisson distribution can be used. This is discussed in Chapter 3. If we use the normal distribution, and if S indicates the number of successes, P the population proportion, $P = 1 - Q$, and N is the sample size, NP is the population number of successes and $\sqrt{NPQ}$ is the standard deviation in the population. We can solve as follows.

For part (a),

$$Z = \frac{S - NP}{\sqrt{NPQ}} = \frac{5 - (10)(0.75)}{\sqrt{10(0.75)(0.25)}} = \frac{-2.5}{1.37} = -1.82.$$

The standardized normal variate for five successes $(S = 5)$ is -1.82 (see Figure 13).

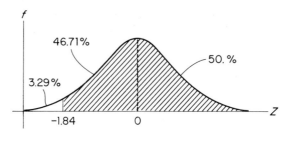

FIG. 13

$$P\left(\frac{S - \mu}{\sigma_s} > -1.82\right) = 0.5 + P\left(-1.82 < \frac{S - \mu}{\sigma_s} < 0\right)$$

$$= 0.5 + 0.4656 = 0.9656.$$

And for part (b),

$$Z = \frac{S - NP}{\sqrt{NPQ}} = \frac{12 - (24)(0.75)}{\sqrt{24(0.75)(0.25)}} = \frac{-6}{2.12} = -2.83,$$

$$P\left(-2.83 < \frac{S - \mu}{\sigma_s}\right) = 0.5 + P\left(-2.83 < \frac{S - \mu}{\sigma_s} < 0\right)$$

$$= 0.5 + 0.4977 = 0.9977.$$

The answer obtained from the normal distribution is "closer" to the correct answer (obtained from the binomial distribution) when the sample size is larger.

Problem 34. Poisson Distribution

Q: A department store employs 1,000 workers, and experience indicates that on the average three workers per year are discharged for theft. What is the probability that the firm will

(a) not discharge any workers?
(b) discharge four or more workers in the coming year?

A: This problem involves the occurrence of a rare event (probability = $3/1,000 = 0.003$) in a large number of observations (1,000). If we assume employee thefts are independent of each other, the Poisson distribution can be used as an approximation to the frequency distribution in this situation. The probability of X occurrences is

$$P(X) = \frac{\lambda^x e^{-\lambda}}{X!},$$

where e is the base of the natural logarithm, λ is the mean (and variance) of the number of occurrences of the rare event, and X is a non-negative integer.

A table for the Poisson distribution is given in Table 7 of the Appendix.

(a) For $\lambda = 3$, $P(X = 0) = 0.0498 \approx 0.05$.

(b) $P(X \geq 4) = 1 - P(X \leq 3)$. Then

$$P(X \leq 3) = P(X = 0) + P(X = 1) + P(X = 2)$$
$$+ P(X = 3)$$
$$= 0.0498 + 0.1494 + 0.2240 + 0.2240$$
$$= 0.6472,$$
$$P(X \geq 4) = 1.000 - 0.6472 = 0.3528 \approx 0.35.$$

There is a 5 percent probability that no workers will be discharged for theft and a 35 percent probability that four or more workers will be discharged for theft in the coming year.

Problem 35. Geometric Distribution

Q: An unemployed worker can canvass only one firm per day, and there is a 10 percent probability of his taking a job on any one inquiry.

    (a) What is the probability that the worker will find a job on the third day of his search?

    (b) What is the probability that he will find a job within the first three days of his search?

    (c) What are the mean and variance of the duration of days of unemployment?

A: (a) The probability of a worker finding a job on the Xth day but not before is $P(X) = (1 - P)^{x-1}(P)$, where P is the probability of his finding a job on a particular day.

$$P(X = 3) = (1.0 - 0.1)^2(0.1) = (0.9)^2(0.1) = 0.081.$$

    (b) The probability of the worker's finding a job in the first three days is

P(finding job in first 3 days)

$$= \sum_{i=1}^{3} P(X_i) = (0.9)^0(0.1) + (0.9)(0.1) + (0.9)^2(0.1)$$

$$= 0.1 + 0.09 + 0.081 = 0.271.$$

(c) This is the geometric distribution, which has a mean $1/P$ and a variance $(1 - P)/P^2$. The mean duration of unemployment is

$$\frac{1}{P} = \frac{1}{0.1} = 10 \text{ days,}$$

and the variance is

$$\frac{1 - P}{P^2} = \frac{1 - 0.1}{(0.1)^2} = \frac{0.9}{0.01} = 90.$$

Chapter 3

Hypothesis Testing: Means and Proportions

This chapter develops problem-solving techniques for testing hypotheses about the mean value of a variable in a population and the proportion of the members of a population with a particular characteristic. For example, we may be interested in the average (mean) income in a population or the proportion of the population that has income over $10,000. The "population" includes all the individuals, or elements, who could, in principle, be questioned or measured for the variable (income) under investigation. Populations can be exceedingly large, even infinite. For instance, the population of centimeters of air on a particular day is infinite. Polling all the elements in an infinite population is obviously impossible. Even the polling of all the elements in a finite population can be costly.† In general, the larger the number of observations drawn from a population, the greater the cost. However, larger samples from a population do produce more reliable results.

If each element in a population has an equal probability of being selected for a sample, the sample is said to be *random*. When samples are randomly selected, statistical theory provides information about the reliability of the sample value (or sample statistic) as an estimate of the population value (or population parameter).

Sample data are used to *test a hypothesis* or to *construct a confidence interval*. A hypothesis is an assumption about a characteristic of a population. On the basis of the hypothesized population value, statistical theory, and a known probability of being wrong, we can construct an *acceptance interval*. If the sample value falls within the acceptance interval, we say that we "accept" the hypothesis as being true, although this statement is subject to a probability that the hypothesis is not true. If the sample value falls outside the acceptance interval, we say that we

---

† The classic example of the prohibitive cost of sampling an entire finite population is quality control sampling of bulbs for flash cameras. Once the item is sampled, it cannot be sold.

"reject" the hypothesis. The probability of rejecting a true hypothesis is subject to control by the investigator and is often called an *alpha* (α) *error* or a *type I error*. The probability of accepting a hypothesis that is false is called a *beta* (β) *error* or *type II error*. If an investigator selects a smaller probability of rejecting a true hypothesis (alpha error), he increases the width of the acceptance interval, and thereby increases the probability of accepting a false hypothesis (beta error). The only way to decrease both the alpha error and the beta error is to increase the size of the sample.

A confidence interval combines the sample data, statistical theory, and alpha error selected by the investigator. One minus the alpha error is the *level of confidence* we have that the true (but unknown) population value falls within the interval.

An important relationship exists between the variance of values in a population (σ_x^2) and the variance of the distribution of sample means ($\sigma_{\bar{x}}^2$). Under random sampling, observations in a sample are independent of each other. Then, since

$$\bar{X} = \frac{\sum\limits_{i=1}^{N} X_i}{N} \, ,$$

it follows that

$$\sigma_{\bar{x}}^2 = \frac{1}{N^2} \left[\sigma^2 (\sum\limits_{i=1}^{N} X_i) \right] = \frac{\sigma_{x_i}^2}{N} \, .$$

The larger the size of the sample drawn from the population, the smaller the variance of the sampling distribution. A smaller sampling variance provides more reliable estimates of the population mean.

Four distributions are used to test hypotheses and construct confidence intervals for the mean value of a continuous variable or the proportion of occurrences of an event in a dichotomous population. They are the normal distribution, the *t*-distribution, the Poisson distribution, and the binomial distribution. The probability distributions for these variables are presented in the Appendix.

If a variable is normally distributed and has a zero mean and a unit standard deviation, it has a *standardized normal distribution*. Suppose the population distribution of a mean ($\bar{X}$) is normally distributed and has a mean $\mu_{\bar{x}}$ and a standard deviation $\sigma_{\bar{x}}$. Then,

$$Z = \frac{\bar{X} - \mu_{\bar{x}}}{\sigma_{\bar{x}}}$$

is a standardized normal variable. Sample means ($\bar{X}$) are normally distributed if the sampling is random and if either the underlying population (X_i) is normally distributed or the sample is large.† Since the mean of the population of sample means ($\mu_{\bar{x}}$) is the same as the mean in the underlying population (μ_x), and since $\sigma_{\bar{x}}^2 = \sigma_x^2/N$, we can write

$$Z = \frac{(\bar{X} - \mu_x)\sqrt{N}}{\sigma_x}.$$

This permits the conversion of data on the sample mean and population values into a standardized normal variate. Table 1 in the Appendix presents the probability distribution for the standardized normal variate.

If the population standard deviation is not known, we cannot compute a standardized normal variate. However, if $\bar{X}$ is normally distributed and the sample standard deviation (S) is known, we can compute

$$t = \frac{\bar{X} - \mu_{\bar{x}}}{S_{\bar{x}}} = \frac{(\bar{X} - \mu_x)\sqrt{N}}{S_x},$$

where the mean value of t is zero, and the standard deviation of t is unity. The *t-distribution* (also called the *Student t-distribution*, after the pseudonym of its developer) is symmetric around the value zero and ap-

---

† The *Central Limit Theorem* states that a linear combination of independent random variables approaches a normal distribution for large samples. The observations in a random sample are independent of each other and a mean is a weighted sum. Hence, regardless of the shape of the distribution of the underlying population (X_i), the distribution of sample means ($\bar{X}$) will approximate a normal distribution for large samples.

proaches the normal distribution for large samples. Table 2 in the Appendix presents the probability distribution of the t-distribution.

The probability distribution of the t-statistic is a function of a parameter called the number of *degrees of freedom* (df). The larger the number of degrees of freedom, the smaller the variance of the t-distribution and the closer the approximation of the t-distribution to the normal distribution. The approximation is very close for 30 or more degrees of freedom. The number of degrees of freedom is the number of "linearly independent observations," that is, the number of randomly selected observations minus the number of constraints placed on the data. For example, in testing hypotheses about means, one constraint is placed on the sample data when the observed sample standard deviation is used rather than the standard deviation in the population. Thus, for tests of hypotheses about means, the t-distribution has $N - 1$ degrees of freedom.

If $\bar{X}$ is not normally distributed (for example, because the sample is small and the underlying population is not normal), neither the normal distribution nor the t-distribution can be used to construct confidence or acceptance intervals. Other procedures (e.g., Chebyshev's inequality, or transformations of the data) may be applied.

Table 1 summarizes rules of thumb for applying the normal distribution and the t-distribution to analyses of means of continuous variables.

If we wish to test a hypothesis or construct a confidence interval for the proportion of successes in a dichotomous population, we can always

TABLE 1

Rules of Thumb for Testing Hypotheses about Means

	Distribution of parent population	
Sample size	Normal	Not normal
Large $(N > 30)$		
Population variance known	Normal distribution	Normal distribution
Population variance unknown	Normal distribution†	Normal distribution†
Small $(N < 30)$		
Population variance known	Normal distribution	Neither normal nor t-distribution can be applied
Population variance unknown	t-distribution	Neither normal nor t-distribution can be applied

† As an approximation to the t-distribution.

use the *binomial distribution* (or binomial expansion) to obtain the probability statements. This is an exceedingly tedious task for samples that are not very small, and under certain circumstances other distributions provide very close approximations to the binomial distribution.

If we let P be the proportion of "successes" in a dichotomous population and N the sample size, we can compute the mean number of successes as NP. The distribution of the proportion of successes (p) among samples, or the sampling distribution of the number of successes (Np), approaches a normal distribution if $NP > 10$ and $N(1 - P) > 10$. Then,

$$t = \frac{p - P}{S_p} = \frac{Np - NP}{S_{Np}}$$

has a t-distribution.

If a sample is size N and the proportion of successes is p, the number of successes can be written as $X = Np$. Taking the standard deviation of both sides of this equation gives $S_x = S_{Np}$. Thus,

$$t = \frac{p - P}{S_p} = \frac{X - \mu_x}{S_x}$$

and a problem of testing a population proportion (P) can be converted into a problem of testing the value of a continuous variable (μ). If the sample size is large $(N > 30)$, the normal distribution can be used as an approximation to the t-distribution.

However, if NP or $N(1 - P)$ is small (say, less than 10), the sampling distribution of the proportion of successes (p) or the number of successes (Np) is not normally distributed. For a large sample and a very small P or $1 - P$ the *Poisson distribution* is a close approximation to the binomial distribution. The Poisson distribution is a function of one parameter, λ, which is the mean and variance of the number of successes.† Tables for the Poisson distribution indicate the probability of obtaining a particular number of successes (X) for different values of the mean number of successes (λ). (See Table 7 in the Appendix.)

† $\lambda = NP \approx NP(1 - P)$ when N is very large and P is close to zero.

TABLE 2
Rules of Thumb for Testing Hypotheses about Proportions
(N = sample size, P = probability of a success in the population, NP = expected number of successes.)

1. $NP > 10$ and $N(1 - P) > 10$. Sample proportion is approximately normally distributed.
   (a) If N is large ($N > 30$), use the normal distribution as an approximation to the t-distribution.
   (b) If N is small ($N < 30$), use the t-distribution.
2. $NP < 10$ or $N(1 - P) < 10$. Sample proportion is *not* normally distributed.
   (a) If N is large and P or $1 - P$ is small, use the Poisson distribution.
   (b) If N is small, use the binomial espansion tables.

If the sample is small and either NP or $N(1 - P)$ is also small, the binomial expansion must be used to obtain probability statements. Tables exist for the values of the binomial expansion for small samples. (See Table 8 in the Appendix.)

Table 2 summarizes rules of thumb for applying the normal, t, Poisson, and binomial distributions to analyses of population proportions.

Problems

1. Confidence Interval: Mean
2. Hypothesis Testing: Mean
3. Hypothesis Testing: Mean
4. Hypothesis Testing: Proportion
5. Hypothesis Testing: Proportion
6. Confidence Interval: Proportion
7. Hypothesis Testing: Proportion, One Tailed Test
8. Hypothesis Testing: Proportion, Rare Events (Poisson Distribution)
9. Hypothesis Testing: Median
10. Confidence Interval Width: Proportion
11. Confidence Interval Width: Mean
12. Confidence Interval Width: Chebyshev's Inequality
13. Hypothesis Testing: Difference in Means, Large Samples
14. Hypothesis Testing: Difference in Means, Small Samples
15. Hypothesis Testing: Matched Samples
16. Sample Size and Type I and Type II Errors

Problem 1. Confidence Interval: Mean

Q: (a) A random sample of 100 households is found to have a mean annual income of $5,000. It is known that the standard deviation of income in the population is $300. With a 95 percent level of confidence, what is the population mean?

(b) What is the confidence interval if the population standard deviation is unknown but the sample standard deviation is $300? Would the width of the confidence interval differ substantially if the normal distribution were used as an approximation to the t-distribution?

A: (a) A sample of more than 30 is sufficiently large for the sample mean to be normally distributed. Since the population variance is known, $(\bar{X} - \mu)/\sigma_{\bar{x}}$ is normally distributed and the normal distribution is appropriate for constructing the confidence interval. The value of the standardized normal variate (Table 1 in the Appendix) which cuts off 2.5 percent of the area at each tail of the distribution is $Z = \pm 1.96$ (see Figure 1).

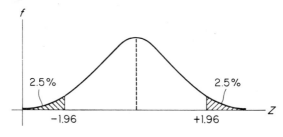

FIG. 1

$$P\left(-1.96 < \frac{\bar{X} - \mu}{\sigma_{\bar{x}}} < +1.96\right) = 0.95,$$

$P(\bar{X} - (1.96)\sigma_{\bar{x}} < \mu < \bar{X} + (1.96)\sigma_{\bar{x}}) = 0.95,$

$\sigma_{\bar{x}} = \dfrac{\sigma_x}{\sqrt{N}} = \dfrac{300}{10} = 30,$

$P(5,000 - 58.8 < \mu < 5,000 + 58.8) = 0.95,$

$P(4,941.2 < \mu < 5058.8) = 0.95.$

There is a probability of 0.95 that the values $4,941.20 to $5,058.80 bound the true value of the population mean.

(b) Since the sample size is large, using the Central Limit Theorem, $\bar{X}$ is normally distributed. Since σ_x is unknown but S_x is known, we can use the t-distribution. The value of the t-distribution that cuts off 2.5 percent of the area at each tail of the distribution depends on the number of degrees of freedom, which equals $N - 1$ in testing hypotheses and constructing confidence intervals for means when the sample standard deviation is used. Using the t-distribution table (Table 2 in the Appendix), we find that $t_{0.025} = -2$ for $df = N - 1 = 99$, where df is degrees of freedom.

$$P\left(t_{0.025} < \frac{\bar{X} - \mu}{S_{\bar{x}}} < t_{0.975}\right) = 0.95,$$

$$P(\bar{X} - t_{0.975}S_{\bar{x}} < \mu < \bar{X} - t_{0.025}S_{\bar{x}}) = 0.95,$$

$$S_{\bar{x}} = \frac{S_x}{\sqrt{N}} = \frac{300}{10} = 30,$$

$$P(5,000 - (2)(30) < \mu < 5,000 + (2)(30)) = 0.95,$$

$$P(4,940 < \mu < 5,060) = 0.95.$$

Note the very small difference in the size of the confidence interval if, for a large sample, the normal distribution were used as an approximation to the t-distribution. However, the confidence interval is wider when we lose information about the true variance in the population.

Problem 2. Hypothesis Testing: Mean

Q: Using a 5 percent probability of being wrong, test the following hypotheses:

(a) Mean income is $6,000.
(b) Mean income is greater than $6,000.

Assume a sample size of 64 and a sample standard deviation of $1,600.

A: (a) The null and alternative hypotheses in part (a) are $H_0:\mu = $6,000$ and $H_a:\mu \neq $6,000$. The critical region (i.e., values of $\bar{X}$ that result in a rejection of the null hypothesis) is in the two tails of the distribution (see Figure 2). The value of the normal

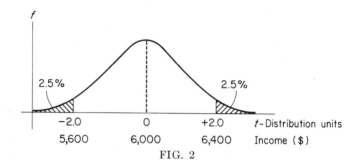

FIG. 2

variate for a type I (or alpha) error of 5 percent is $t = \pm 2$ $(df = N - 1 = 63)$. If the null hypothesis is true, then

$$P\left(t_{\alpha/2} < \frac{\bar{X} - \mu}{S_{\bar{x}}} < t_{1-\alpha/2}\right) = 1 - \alpha,$$

$$P\left(-2 < \frac{\bar{X} - 6,000}{200} < +2\right) = 0.95 \qquad \left[S_{\bar{x}} = \frac{S_x}{\sqrt{N}} = 200\right],$$

$$P(5,600 < \bar{X} < 6,400) = 0.95.$$

The null hypothesis that $\mu = $6,000$ is accepted if the sample mean is in the interval $5,600 to $6,400. There is, however, a 5 percent probability that the null hypothesis is correct even if the sample mean falls outside the interval.

  (b) The null hypothesis is $H_0:\mu \geq $6,000$ and the alternative hypothesis is $H_a:\mu < $6,000$. The null hypothesis is rejected only if the sample mean falls significantly below $6,000. This involves a one-tailed test (left tail), and for a probability of 0.05 of being wrong in rejecting the null hypothesis, the critical value is $t = 1.67$ $(df = N - 1 = 63)$ (see Figure 3).

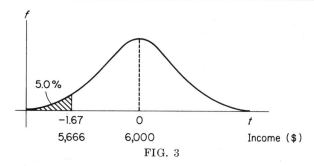

FIG. 3

$$P\left(t_\alpha < \frac{X - \mu}{S_{\bar{x}}}\right) = 1 - \alpha,$$

$$P\left(-1.67 < \frac{\bar{X} - 6{,}000}{200}\right) = 0.95,$$

$$P(5{,}666 < \bar{X}) = 0.95.$$

If the null hypothesis is true, 95 percent of random samples of size 64 will have a mean greater than \$5,666. The null hypothesis is rejected if the sample mean falls below this value.

Note: Rather than writing the acceptance interval, a common procedure is to compute the observed value of the standardized normal variable (or *t*-ratio) and compare it with the critical value. Following this procedure, if the sample mean is \$5,800, the answers to Problem 2 are:

(a) Observed *t*-value

$$t_o = \frac{\bar{X} - \mu}{\sigma_{\bar{x}}} = \frac{5{,}800 - 6{,}000}{200} = -1.0.$$

The critical value is $t_c = \pm 2$. The null hypothesis is not rejected.

(b) Observed *t*-value

$$t_o = \frac{\bar{X} - \mu}{\sigma_{\bar{x}}} = \frac{5{,}800 - 6{,}000}{200} = -1.0.$$

The critical value is $t_c = -1.67$. The null hypothesis is not rejected.

The null hypothesis would have been rejected if the observed t-ratio (t_o) had been larger than the critical t-ratio (t_c).

Problem 3. Hypothesis Testing: Mean

Q: It is hypothesized that the mean number of years of schooling in a population is 15. Six persons are selected randomly from this population, which is known to be normally distributed. Test the hypothesis given that the sample mean is 12.3 and that

(a) the population variance is known to be $\sigma_x^2 = 4.8$,
(b) the population variance is unknown, but the sample variance is 4.8.

A: $H_0 : \mu = 15$ and $H_a : \mu \neq 15$. The type I error (area of the critical region) is not specified, but the value $\alpha = 0.01 = 1$ percent will be used. Since the population is normally distributed the sample mean ($\bar{X}$) is normally distributed even though the sample size is small.

(a) Since the population variance (σ^2) is known, $(\bar{X} - \mu)/\sigma_{\bar{x}}$ is normally distributed. The value of the standardized normal variable that cuts off 0.5 percent of the area under the normal curve at each tail is $Z = \pm 2.58$ (see Figure 4). The acceptance interval is

$$P\left(Z_{\alpha/2} < \frac{\bar{X} - \mu}{\sigma_{\bar{x}}} < Z_{1-\alpha/2}\right) = 1 - \alpha,$$

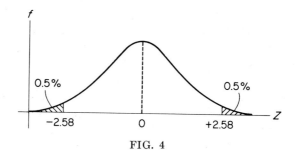

FIG. 4

$$\sigma_{\bar{x}}^2 = \frac{\sigma^2}{N} = \frac{4.8}{6} = 0.8, \qquad \sigma_{\bar{x}} \approx 0.9,$$

$$P\big(15 - (2.58)(0.9) < \bar{X} < 15 + (2.58)(0.9)\big) = 0.99,$$

$$P(12.7 < \bar{X} < 17.3) = 0.99.$$

Since the observed $\bar{X} = 12.3$, the null hypothesis that $\mu = 15$ is rejected.

(b) Since the population variance is unknown but the sample variance is known, $(\bar{X} - \mu)/S_{\bar{x}}$ can be computed. It has a t-distribution because the population has a normal distribution.

$$S_{\bar{x}} = \frac{S_x}{\sqrt{N}} \approx 0.9,$$

$$P\left(t_{\alpha/2} < \frac{\bar{X} - \mu}{S_{\bar{x}}} < t_{1-\alpha/2}\right) = 1 - \alpha.$$

For six observations, there are only five degrees of freedom ($df = N - 1$). The t-ratio is $t = \pm 4.03$ for a two-tailed test, $\alpha = 0.01$, $df = 5$.

$$P\left(-4.03 < \frac{\bar{X} - 15}{0.9} < 4.03\right) = 0.99,$$

$$P(11.37 < \bar{X} < 18.63) = 0.99.$$

The null hypothesis is accepted. Note the widening of the acceptance interval as we lose information about the true population variance.

Note: For small samples, if the population is not normally distributed neither of the above procedures can be used. One solution is to transform the data to obtain a population distribution that is approximately normal. (For example, if the logarithm of the variable in the population has a normal distribution, taking the antilog of the variable will result in a normally distributed transformed population.) Another solution is to use Chebyshev's Inequality (see Chapter 2).

Problem 4. Hypothesis Testing: Proportion

Q: Using a random sample of 64 observations, test the hypothesis that 40 percent of college students are married.

A: The hypothesized population proportion is $P = 0.4$. The sample proportion has a mean P and a standard deviation $\sqrt{PQ/N}$ if the null hypothesis is correct. Since the sample is large and both NP and $N(1 - P)$ are greater than 10, the binomial distribution approaches the normal distribution. For a type I error of 5 percent,

$$P\left(Z_{\alpha/2} < \frac{p - P}{\sigma_p} < Z_{1-\alpha/2}\right) = 1 - \alpha,$$

$$P\left(-1.96 < \frac{p - 0.4}{0.06} < +1.96\right) = 0.95$$

since

$$\sigma_p = \sqrt{\frac{PQ}{N}} = \sqrt{\frac{(0.4)(0.6)}{64}} = 0.06.$$

Therefore

$$P(0.28 < p < 0.52) = 0.95.$$

Sample proportions in the interval 0.28 to 0.52 result in the acceptance of the null hypothesis ($H_0: P = 0.4$).

Problem 5. Hypothesis Testing: Proportion

Q: A firm knows that 20 percent of its trainees do not successfully complete the program. Last year 60 percent of the group of 100 in one training school were failures. Was this group different from the population, or could the result be caused by random sampling?

A: Here sample proportion $p = 0.6$, population proportion $P = 0.2$, and sample size $N = 100$, so $NP = (100)(0.2) = 20$. Thus the normal approximation can be used.

$$Z_p = \frac{p - P}{\sigma_p} = \frac{p - P}{\sqrt{PQ/N}} = \frac{0.6 - 0.2}{\sqrt{(0.2)(0.8)/100}} = 10$$

or

$$\boxed{Z_x = \frac{X - NP}{\sigma_x} = \frac{Np - NP}{\sqrt{NPQ}} = \frac{p - P}{\sqrt{PQ/N}} = Z_p.}$$

The probability that a normally distributed variable will exceed the population proportion (or the expected number of successes NP) by at least ten standard deviations is approximately zero. We reject the hypothesis that the difference is the result of random sampling.

Problem 6. Confidence Interval: Proportion

Q: In a random sample of 64 households, 70 percent had two or more television sets. With a 95 percent level of confidence, what is the population proportion? The data indicate that the normal approximation is appropriate.

A: The confidence interval is:

$$P\left(Z_{\alpha/2} < \frac{p - P}{\sigma_p} < Z_{1-\alpha/2}\right) = 1 - \alpha,$$

$$P(p - Z_{1-\alpha/2}\sigma_p < P < p - Z_{\alpha/2}\sigma_p) = 1 - \alpha,$$

$$P(0.7 - 1.96\sigma_p < P < 0.7 + 1.96\sigma_p) = 0.95.$$

An estimate is needed for the standard deviation of the proportion.

(a) Conservative approach: Pick the largest σ_p, which gives the widest confidence interval ($p = 0.5$, $q = 0.5$ and $pq = 0.25$):

$$\sigma_p = \sqrt{\frac{0.25}{N}}.$$

(b) Optimistic approach: Use sample proportion:

$$\sigma_p = \sqrt{\frac{pq}{N}}.$$

For this problem $\sigma_p = \sqrt{0.25/64} = 0.063$ by the conservative approach and $\sigma_p = \sqrt{(0.7)(0.3)/64} = 0.057$ by the optimistic approach. The conservative approach is preferable.

$P(0.577 < P < 0.823) = 0.95$.

With a 95 percent level of confidence, the population proportion is in the interval bounded by 0.577 and 0.823.

Note: The width of the acceptance interval is greater when the conservative estimate rather than the optimistic estimate of the standard deviation of the sample proportion is used.

Problem 7. Hypothesis Testing: Proportion, One-Tailed Test

Q: In the past, 60 percent of an entering class in a particular graduate school received their Ph.D.'s within five years. For the 1967 entering class of 100, 67 percent received their degrees within five years. Did this class perform better than previous ones?

A: This is a one-tailed test, so we set up the null hypothesis that the population proportion is less than or equal to 0.6. $H_0: P \leq 0.6$. The alternative hypothesis is then $H_a: P > 0.6$. For a critical area of 5 percent, $Z = 1.64$ (see Figure 5).

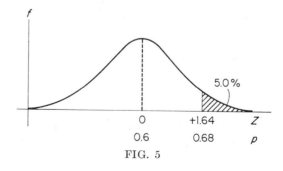

FIG. 5

$$\sigma_p = \sqrt{\frac{(0.6)\,(0.4)}{100}} \approx 0.05.$$

$$\boxed{P\left(\frac{p-P}{\sigma_p} < Z_{1-\alpha}\right) = 1 - \alpha,}$$

$$P(p < P + Z\sigma_p) = 1 - \alpha,$$

$$P(p < 0.68) = 0.95.$$

Since $p = 0.67$, the null hypothesis is not rejected; that is, we cannot reject the hypothesis that this class did not perform better than previous classes.

An alternative procedure is to compare the critical Z-value and the observed Z-value:

$$Z_o = \frac{p-P}{\sigma_p} = \frac{0.67 - 0.60}{\sqrt{(0.6)\,(0.4)/100}} = \frac{0.07}{0.05} = 1.40,$$

$$Z_c = 1.64.$$

Since the observed value of the standardized normal variable is less than the critical value, the null hypothesis cannot be rejected.

Problem 8. Hypothesis Testing: Proportion, Rare Events
(Poisson Distribution)

Q: An insurance company has 4,000 policyholders. What is the probability that not more than three will file claims in a year if it is hypothesized that

(a) the probability of a claim being filed is 0.001?
(b) the probability of a claim being filed is 0.01?

A: (a) $NP = (4,000)\,(0.001) = 4$ is small (i.e., less than 10), so the normal approximation to the binomial does not apply. The Poisson distribution approximation to the binomial distribution holds in this case.

$$P(S \leq 3) = P(S = 0) + P(S = 1) + P(S = 2)$$
$$+ P(S = 3).$$

$$P(S = X) = \binom{N}{X} P^x (1 - P)^{N-X} \approx e^{-\lambda} \frac{\lambda^x}{X!},$$

$$\underbrace{\qquad\qquad\qquad}_{\text{Binomial}} \quad \underbrace{\qquad}_{\text{Poisson}}$$

where $\lambda = NP = $ the mean of the distribution (see Figure 6).

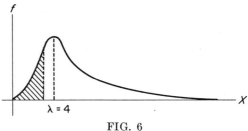

FIG. 6

The approximation holds only when N is large, P is small, and NP is small.† The answer $P(S \leq 3) = 0.4335$ can be found by direct computation, or by using a Poisson distribution table (Appendix Table 7).

Since the variance of outcomes (S) of a binomial distribution is Var $(S) = NPQ$, when P is small $Q = 1 - P \approx 1.0$ and Var $(S) \approx NP = \lambda$. The mean and variance of the Poisson distribution are $\lambda = NP$.

(b) Although N is large and P is small, $NP = (4,000)(0.01) = 40$, which is greater than 10, so the normal approximation to the binomial is appropriate:

$$Z = \frac{S - NP}{\sqrt{NPQ}} = \frac{3 - 400}{\sqrt{400}} = -87.8.$$

† See, for example, Paul G. Hoel, *Introduction to Mathematical Statistics*, 2nd ed. (John Wiley, New York, 1954), pp. 68–69. e is the base of the system of natural logarithms; $e \approx 2.71828$.

The probability that not more than three will file claims is approximately zero.

Problem 9. Hypothesis Testing: Median

Q: Median family income is claimed to be \$7,000. In a random sample of 100 families 23 have incomes above this level.

(a) Do you believe the claim?
(b) With 95 percent confidence, what would you assert to be the upper limit to the proportion of families with incomes over \$7,000, given the sample information?

A: (a) By the definition of median, the hypothesis is that half of the families have incomes greater than \$7,000: $H_0: P = 0.5$, $H_a: P \neq 0.5$, where P is the proportion of families with incomes above \$7,000. For $\alpha = 0.05$, the observed standard normal variate is

$$Z = \frac{p - P}{\sigma_p},$$

where

$$\sigma_p = \sqrt{\frac{(0.5)(0.5)}{100}} = 0.05, \qquad Z = \frac{0.23 - 0.5}{0.05} = -5.4.$$

The probability of obtaining an observation more than five standard deviations below the mean is approximately zero. The hypothesis is rejected.

(b) $P(P \leq p + Z\sigma_p) = 1 - \alpha,$

$P(P \leq 0.23 + (1.64)(0.05)) = 0.95,$

$P(P \leq 0.312) = 0.95.$

There is a 95 percent probability that the population proportion is less than the value 0.312 when the conservative estimate of the variance of the sampling proportion is used (see Figure 7).

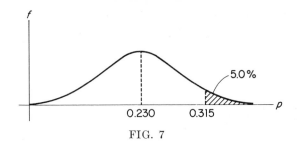

FIG. 7

Problem 10. Confidence Interval Width: Proportion

Q: A researcher is interested in constructing a confidence interval for the proportion of male workers in the population who belong to unions. He wants the confidence interval to be six percentage points wide and he wants a 95 percent level of confidence. How large a random sample does he want?

A: If he believes the sample would be sufficiently large, the normal distribution is appropriate. Then, the confidence interval is

$$P\left(-Z < \frac{p - P}{\sigma_p} < +Z\right) = 1 - \alpha,$$

$$P(p - Z\sigma_p < P < p + Z\sigma_p) = 1 - \alpha.$$

The width of the confidence interval is $2Z_{\alpha/2}\sigma_p$. Since no sample values are available, to be conservative the largest σ_p is

$$\sqrt{\frac{(0.5)(0.5)}{N}} = \frac{0.5}{\sqrt{N}}.$$

Then, $0.06 = 2(1.96)(0.5/\sqrt{N})$ or $N \approx 1{,}067$, which is large. Hence the procedure of assuming a normal distribution is appropriate.

Problem 11. Confidence Interval Width: Mean

Q: How large a random sample is needed for a confidence interval of $500 for the mean income in a population if a 95 percent level of confidence is desired and the standard deviation in a sample is $2,500?

A: If income is normally distributed in the population or if the sample size is expected to be large, the normal distribution may be used. The width of the interval is

$$w = 2t_{\alpha/2}S_{\bar{x}} = 2t_{\alpha/2}\frac{S_x}{\sqrt{N}}.$$

If it is assumed the sample size will be large, $t_{\alpha/2} = 1.96$. Then

$$500 = (2)(1.96)\frac{2{,}500}{\sqrt{N}},$$

$$N = 384.$$

A large sample of 384 observations should be selected.

Problem 12. Confidence Interval Width: Chebyshev's Inequality

Q: If a population has a standard deviation of $2,000, how large a sample (N) is needed to have at least 95 percent confidence that a sample mean is within $100 of the (unknown) population mean?

A: If it is not certain that the sample will be large, and if the shape of the population is not known, we use Chebyshev's Inequality (see Chapter 2), and the relation $\sigma_{\bar{x}}^2 = \sigma_x^2/N$.

$$P\left(\frac{|\bar{X} - \mu|}{\sigma_{\bar{x}}} < h\right) \geq 1 - \frac{1}{h^2}.$$

If $h\sigma_{\bar{x}} = 100$, then

$$P(|\bar{X} - \mu| < 100) \geq 1 - \frac{1}{(100/\sigma_{\bar{x}})^2} = 1 - \frac{\sigma_{\bar{x}}^2}{(100)^2},$$

$$0.95 = 1 - \frac{\sigma_{\bar{x}}^2}{(100)^2} = 1 - \frac{\sigma_x^2}{N(100)^2},$$

$$N = \left(\frac{1}{0.05}\right)\left(\frac{2{,}000}{100}\right)^2 = 8{,}000.$$

Therefore we know the sample is large. Using the Central Limit Theorem, for large samples, regardless of the distribution of X_i, linear combinations of X_i are normally distributed. Hence the distribution of sample means is normal. Then, using this information, we have

$$P(-100 < \bar{X} - \mu < 100) \geq 0.95,$$

$$P\left(\frac{-100}{\sigma_{\bar{x}}} < \frac{X - \mu}{\sigma_{\bar{x}}} < \frac{100}{\sigma_{\bar{x}}}\right) \geq 0.95,$$

$$Z_{0.025} = 1.96 = \frac{100}{\sigma_{\bar{x}}} = \frac{\sqrt{N}(100)}{\sigma_x},$$

$$N = \frac{(1.96)^2 \sigma_x^2}{(100)^2} = (40)^2 = 1,600.$$

Note: Notice the power of the Central Limit Theorem. Knowing that we can use a normal distribution reduces the sample size from 8,000 to 1,600 observations.

Problem 13. Hypothesis Testing: Difference in Means, Large Samples

Q: A study of the weekly income (X) of construction workers in New York and Los Angeles yields the following figures:

New York	Los Angeles
$\bar{X}_1 = \$250$	$\bar{X}_2 = \$240$
$S_1^2 = \quad 400$	$S_2^2 = \quad 360$
$N_1 = \quad 40$	$N_2 = \quad 60$

Assuming random samples, does income differ in the two regions?

A: The null hypothesis is $H_0: \mu_1 = \mu_2$ and the alternative hypothesis is $H_a: \mu_1 \neq \mu_2$. The samples are independent and are sufficiently large for $\bar{d} = \bar{X}_1 - \bar{X}_2$ to be approximately normally distributed. The variance of the difference in sample means is

$$S_{\bar{d}}^2 = S_{\bar{x}_1}^2 + S_{\bar{x}_2}^2 - 2 \, \text{Cov} \, (\bar{X}_1, \bar{X}_2),$$

but the covariance term equals zero since the samples are independent. Therefore

$$S_{\bar{d}} = \sqrt{S_{\bar{x}_1}{}^2 + S_{\bar{x}_2}{}^2} = \sqrt{\frac{S_1{}^2}{N_1} + \frac{S_2{}^2}{N_2}} = \sqrt{\frac{400}{40} + \frac{360}{60}},$$

$S_{\bar{d}} = 4.0$

Hence

$$P\left(t_{\alpha/2} < \frac{\bar{d} - D}{S_{\bar{d}}} < t_{1-\alpha/2}\right) = 1 - \alpha,$$

where $D = \mu_1 - \mu_2 = 0$. There are $N_1 + N_2 - 2 = 98$ degrees of freedom. For $\alpha = 0.05$, $t_{\alpha/2} = 1.98$.

$$P(D + S_{\bar{d}}t_{\alpha/2} < \bar{d} < D + S_{\bar{d}}t_{1-\alpha/2}) = 1 - \alpha.$$

If $D = 0$, then

$$P(-7.92 < \bar{d} < +7.92) = 0.95.$$

The observed difference $\bar{d} = \$10$ falls outside the acceptance interval, and the null hypothesis is rejected (see Figure 8).

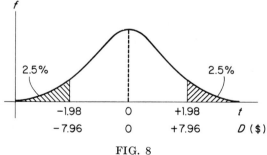

FIG. 8

Problem 14. Hypothesis Testing: Difference in Means, Small Samples

Q: Two independent samples were taken of the weight (X) of NFL and AFL football players:

AFL	NFL
$\bar{X}_1 = 250$	$\bar{X}_2 = 240$
$S_1^2 = 400$	$S_2^2 = 360$
$N_1 = \ \ 8$	$N_2 = \ \ 12$

Is there a difference in weight in the two leagues?

A: The samples are small. If the population is normally distributed, $\bar{d} = \bar{X}_1 - \bar{X}_2$ is normally distributed. If the population variances were known, the normal distribution would be applied. However, since only the sample variances are known, the t-distribution can be used. The number of degrees of freedom is $df = N_1 + N_2 - 2$ only if the population variances do not differ. Since the sample variances are approximately equal, the population variances are assumed not to differ.†

The null and alternative hypotheses are

$$H_0: D = \mu_1 - \mu_2 = 0, \qquad H_a: D = \mu_1 - \mu_2 \neq 0.$$

$$P\left(t_{\alpha/2} < \frac{\bar{d} - D}{S_{\bar{d}}} < t_{1-\alpha/2}\right) = 1 - \alpha,$$

$$S_{\bar{d}} = \sqrt{\frac{S_1^2}{N_1} + \frac{S_2^2}{N_2}} = \sqrt{\frac{400}{8} + \frac{360}{12}} = \sqrt{80} \approx 9.$$

† The number of degrees of freedom is smaller if the two populations have significantly different variances. See, for example, H. Walker and J. Lev, *Statistical Inference* (Holt, Rhinehart and Winston, New York, 1953), p. 158. The statistical testing of the hypothesis of equality of variances is developed in Chapter 4.

For $df = N_1 + N_2 - 2 = 18$ and $\alpha = 0.05$, $t_{\alpha/2} = 2.10$. The acceptance interval is

$$D \pm t_{\alpha/2}S_{\bar{d}} = 0 \pm (2.1)(9) = \pm18.9.$$

The observed difference of 10 falls within the acceptance interval. The null hypothesis of no difference in weight is not rejected.

Note: For Problems 13 and 14, if it is known that $\sigma_1^2 = \sigma_2^2$, the sample standard deviations can be pooled:

$$\hat{S}^2 = \frac{\sum\limits^{N_1} (X_1 - \bar{X}_1)^2 + \sum\limits^{N_2} (X_2 - \bar{X}_2)^2}{N_1 + N_2 - 2}$$

$$= \frac{(N_1 - 1)S_1^2 + (N_2 - 1)S_2^2}{N_1 + N_2 - 2}.$$

When $\sigma_1^2 = \sigma_2^2$, $\hat{S}^2$ is an unbiased estimate of σ^2. Then

$$S_{\bar{d}} = \hat{S}\sqrt{\frac{1}{N_1} + \frac{1}{N_2}}.$$

Problem 15. Hypothesis Testing: Matched Samples

Q: One hundred families were sampled in the first week of 1971 and again in 1972. The mean level of weekly income was $100 and $102, respectively. Construct an acceptance interval and test the hypotheses that

(a) the mean weekly income did not change and
(b) weekly wages increased by $3.00.

The standard deviation of the change in income in the sample is $3.00.

A: (a) $H_0: D = \mu_{1972} - \mu_{1971} = 0$, $H_a: D = \mu_{1972} - \mu_{1971} \neq 0$. The data can be viewed as *one* sample of differentials (d_i). The acceptance interval is $D \pm t_{\alpha/2}S_{\bar{d}}$, where there are $N - 1 = 99$ degrees of freedom. Then, for a 5 percent type I error (or level of significance), $t = 1.98$.

$S_{\bar{d}} = S_d/\sqrt{N}$, where S_d is the standard deviation of the differences in observations from 1971 to 1972. If $S_d = 3$ and $S_{\bar{d}} = \frac{3}{10} = 0.3$, then

$$D \pm t_{\alpha/2}S_{\bar{d}} = 0 \pm (1.98)(0.3) = \pm 0.59.$$

The observed average differential ($\bar{d} = 2.0$) falls outside the acceptance interval, so the null hypothesis is rejected.

(b) $H_0 : D = \mu_{1972} - \mu_{1971} = 3$, $H_a : D = \mu_{1972} - \mu_{1971} \neq 3$.

$$D \pm t_{\alpha/2}S_{\bar{d}} = 3 \pm (1.98)(0.3) = 3.0 \pm 0.59.$$

The null hypothesis is rejected.

Note: The choice between a matched sample and two independent samples depends on the cost of sampling and the correlation between observations in the first and second members of the pair.

For a study of differences, $d = X_1 - X_2$ and

$$\sigma_{\bar{d}}^2 = \frac{1}{N}\,\sigma_d^2 = \frac{1}{N}\left[\sigma_{x_1}^2 + \sigma_{x_2}^2 - 2\,\mathrm{Cov}\,(X_1, X_2)\right]$$

$$= \frac{1}{N}\,(\sigma_{x_1}^2 + \sigma_{x_2}^2 - 2R_{12}\sigma_{x_1}\sigma_{x_2}),$$

where R_{12} is the correlation coefficient between X_1 and X_2.

In independent samples R_{12} equals zero. If the correlation between the members of the matched pair is positive ($R_{12} > 0$), the variance of sample mean differences ($\sigma_{\bar{d}}^2$) is smaller for the matched sample. Thus if incomes in successive periods are positively correlated, and we are doing a study of intertemporal differences in incomes ($\bar{X}_1 - \bar{X}_2$), matched samples are more efficient statistically than independent samples of the same size.

For a study of sums (e.g., adding husband's income to wife's income to obtain family income), where the elements in the pair are positively correlated, the sampling distribution of mean sums ($\bar{X}_1 + \bar{X}_2$) is smaller for two independent samples than for a matched sample:

$$\sigma_{\overline{x_1+x_2}} = \sigma_{\bar{x}_1}^2 + \sigma_{\bar{x}_2}^2 + 2R_{12}\sigma_{\bar{x}_1}\sigma_{\bar{x}_2},$$

where $R_{12} = 0$ in independent samples.

Thus for a study of sums, where the elements are positively correlated, two independent samples each of size N are more efficient *statistically* than N pairs of matched observations. However, sampling costs are generally lower if $2N$ matched individuals are polled rather than $2N$ independent individuals.

If the correlation between the members of a matched pair is negative $(R_{12} < 0)$, matched samples are statistically more efficient for sums, and independent samples are statistically more efficient for differences (see Table 3).

TABLE 3
Statistical Efficiency

Correlation between members in the matched pair	Study of	
	Sums	Differences
Positive	Independent samples	Matched samples
Negative	Matched samples	Independent samples

Problem 16. Sample Size and Type I and Type II Errors

Q: A high school administrator wants to test the hypothesis that at most 60 percent of 17-year-olds want to go to college. A colleague suggests that 80 percent want to go to college. The administrator wants to have a type I error of 5 percent; that is, he wants a 5 percent probability that he will reject the hypothesis $P = 0.60$ when it is in fact the true hypothesis. He is willing to be more generous to his colleague, and wants a 2.5 percent probability that he will reject the hypothesis that $P = 0.80$ when it is in fact the true hypothesis. What sample size should the administrator use?

A: The administrator's null hypothesis is $H_0: P \leq 0.60$, and the alternative hypothesis is $H_a: P \geq 0.80$.

$$Z = \frac{p - P}{\sqrt{PQ/N}}, \qquad \text{where } Q = 1 - P,$$

$$Z_\alpha = 1.64 = \frac{p - 0.6}{\sqrt{(0.6)(0.4)/N}},$$

$$Z_\beta = -1.96 = \frac{p - 0.8}{\sqrt{(0.8)(0.2)/N}}.$$

There are two equations and two unknowns, sample size (N) and the proportion (p) where the two critical regions meet (see Figure 9). Solving the two equations for N and p, we obtain $N = 63$.

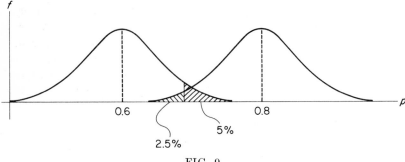

FIG. 9

The administrator should take a random sample of 63 observations.

Note: Given the two hypotheses H_0 and H_a, if we decide on the type I error (α) and the type II error (β), we have determined the sample size. If, however, we select the sample size and the type I error, the critical value of p and the type II error are both determined.

For a fixed sample size, a decrease in the type I error increases the type II error, and vice versa. Both types of error can be decreased by increasing the size of the sample.

Chapter 4

Hypothesis Testing: Variances and Goodness of Fit

This chapter presents problem-solving techniques for testing hypotheses and constructing confidence intervals for the variance in a population. The chi-square (χ^2) distribution, the normal distribution, Fisher's F-distribution, and Hartley's $F_{\max}$ distribution are used here for these purposes. Techniques are also presented for testing whether several samples are drawn from the same population and whether one or more sample distributions come from a hypothesized population. These tests involve the χ^2-distribution. Those who are not familiar with the principles of hypothesis testing are advised to read the introduction to Chapter 3.

Recall that if a variable comes from a normally distributed population with a zero mean and a unit standard deviation, the sample mean (i.e., the sum of N observations divided by N) is itself normally distributed. The sum of the squares of a standardized normally distributed variable obtained from random sampling has a chi-square (χ^2) distribution. By the Central Limit Theorem, the χ^2-distribution approaches the normal distribution for large samples. The variable

$$\sum_{i=1}^{N} \frac{(X_i - \mu)^2}{\sigma^2}$$

has a χ^2-distribution with N degrees of freedom because there are no constraints on the sample values. However,†

† Note that $\sum_{i=1}^{N} [(X_i - \bar{X}) + (\bar{X} - \mu)]^2 = \sum_{i=1}^{N} (X_i - \bar{X})^2 + \sum_{i=1}^{N} (\bar{X} - \mu)^2 +$

$2 \sum_{i=1}^{N} (X_i - \bar{X})(\bar{X} - \mu)$, and $\sum_{i=1}^{N} (X_i - \bar{X}) = 0$. Also,

$$\frac{N(\bar{X} - \mu)^2}{\sigma^2} = \frac{(\bar{X} - \mu)^2}{\sigma^2/N} = \frac{(\bar{X} - \mu)^2}{\sigma_{\bar{x}}^2} = \chi_1^2.$$

$$\chi_N^2 = \sum_{i=1}^{N} \frac{(X_i - \mu)^2}{\sigma^2}$$

$$= \sum_{i=1}^{N} \frac{(X_i - \bar{X} + \bar{X} - \mu)^2}{\sigma^2}$$

$$= \sum_{i=1}^{N} \frac{(X_i - \bar{X})^2}{\sigma^2} + \sum_{i=1}^{N} \frac{(\bar{X} - \mu)^2}{\sigma^2}$$

$$= \sum_{i=1}^{N} \frac{(X_i - \bar{X})^2}{\sigma^2} + \frac{N(\bar{X} - \mu)^2}{\sigma^2}$$

$$= \chi_{N-1}^2 + \chi_1^2.$$

The term

$$\chi^2 = \sum_{i=1}^{N} \frac{(X - \bar{X})^2}{\sigma^2}$$

has $N - 1$ degrees of freedom because one degree of freedom is lost in computing the mean $(\bar{X})$. Since the unbiased estimate of the population variance is

$$S^2 = \frac{\sum_{i=1}^{N} (X_i - \bar{X})^2}{N - 1},$$

it follows that

$$\chi_{N-1}^2 = \sum_{i=1}^{N} \frac{(X_i - \bar{X})^2}{\sigma^2} = \frac{(N - 1)S^2}{\sigma^2}.$$

Thus if we know the null hypothesis of the population variance, the observed sample variance, and the sample size, we can compute the probability of obtaining the χ^2-value or a larger value. We then compare this to the critical value to test the null hypothesis. Similarly, given an alpha or type I error and the sample values, we can compute a confidence interval for the population variance. Table 5 in the Appendix presents the probability distribution for the χ^2-distribution.

The ratio of two χ^2-distributions drawn from independent samples divided by the number of degrees of freedom has a *Fisher F-distribution*:

$$F_{(\eta_1,\eta_2)} = \frac{\chi_1^2/\eta_1}{\chi_2^2/\eta_2} = \frac{S_1^2}{S_2^2}\frac{\sigma_2^2}{\sigma_1^2},$$

where η_1 and η_2 are the degrees of freedom that define the F-distribution.† The F-distribution (Appendix Table 3) is used to test hypotheses (or construct confidence intervals) about the equality or the proportionality of two population variances. For tests of the equality of population variances for more than two samples, *Hartley's* $F_{\max}$ *test* (Appendix Table 6) or *Bartlett's test*‡ can be applied. Table 1 presents rules of thumb for the appropriate distribution when testing hypotheses or constructing confidence intervals for population variances.

The χ^2-distribution is also used to test the *goodness of fit* of an observed distribution (or distributions) to a hypothesized distribution. If f_i is the observed absolute frequency and F_i is the hypothesized absolute fre-

TABLE 1

Rules of Thumb for Testing Hypotheses about Population Variances

---

A. One variance
    1. Small sample: χ^2-distribution
    2. Large sample: Normal distribution as an approximation to the χ^2-distribution
B. Equality or ratio of two variances (from independent samples): F-distribution
C. Equality of more than two variances (from independent samples)
    1. Equal sample sizes: $F_{\max}$ distribution
    2. Unequal sample sizes
       (a) $F_{\max}$ distribution if it gives unambiguous results
       (b) Bartlett's test

---

† Note that $F_{(\eta_1,\eta_2)} = \left(\dfrac{S_1^2}{S_2^2}\right)\left(\dfrac{\sigma_2^2}{\sigma_1^2}\right) = \dfrac{1}{(S_2^2/S_1^2)(\sigma_1^2/\sigma_2^2)} = \dfrac{1}{F_{(\eta_2,\eta_1)}}$.

‡ For a discussion of Bartlett's test, see H. Walker and J. Lev, *Statistical Inference* (Holt, Rinehart and Winston, New York, 1953), p. 193.

quency, then

$$\chi^2_{k-l-1} = \sum_{i=1}^{k} \frac{(f_i - F_i)^2}{F_i}$$

is a χ^2-variable with $k - l - 1$ degrees of freedom. The parameter k is the number of intervals in the sample data, and l is the number of degrees of freedom that are lost when the hypothesized frequencies are computed. For example, suppose an observed distribution is to be tested to see whether it came from a normal distribution, but the population mean and variance of the normal distribution are unknown. Then the sample mean and variance are used as estimates of the population values in the computation of the hypothesized frequencies, and $l = 2$.

A simple proof that the χ^2-distribution can be used to test the goodness of fit is as follows.

Suppose we wish to test whether an observed sample proportion of successes (p) from a large random sample of size N is consistent with a population proportion P_0. Then, since $Q_0 = 1 - P_0$,

$$\chi^2 = \sum_{i=1}^{2} \frac{(f_i - F_i)^2}{F_i} = \frac{(Np - NP_0)^2}{NP_0} + \frac{[N(1-p) - N(1-P_0)]^2}{N(1-P_0)}$$

$$= \frac{(Np - NP_0)^2(1 - P_0) + P_0(Np - NP_0)^2}{NP_0(1 - P_0)}$$

$$= \frac{(Np - NP_0)^2(P_0 + 1 - P_0)}{NP_0(1 - P_0)} = \left(\frac{Np - NP_0}{\sqrt{NP_0Q_0}}\right)^2$$

$$= \frac{(\bar{X} - \mu)^2}{\sigma_x^2} = \chi_1^2.$$

The number of degrees of freedom is $df = k - l - 1 = 1$, since $k = 2$ and $l = 0$.

Problems

1. Hypothesis Testing and Confidence Interval: Variance, Small Sample
2. Hypothesis Testing and Confidence Interval: Variance, Large Sample

3. Hypothesis Testing: Equality of Two Variances
4. Confidence Interval: Ratio of Two Variances
5. Hypothesis Testing: Equality of Several Variances, Equal Sample Size
6. Hypothesis Testing: Equality of Several Variances, Unequal Sample Size
7. Testing Goodness of Fit to a Hypothesized Distribution
8. Testing whether Several Samples Come from the Same Population (Homogeneous χ^2)
9. Testing Goodness of Fit of Several Large Samples to a Hypothesized Distribution (Combined χ^2)
10. Testing Goodness of Fit of Pooled Data to a Hypothesized Distribution (Pooled χ^2)
11. Testing Goodness of Fit of Two or More Small Samples to a Hypothesized Distribution

Problem 1. Hypothesis Testing and Confidence Interval: Variance, Small Sample

Q: A firm is interested in the age distribution of its workers. A random sample of 17 workers is to be used to test a manager's belief that the variance in age is 64. The sample variance is 100.

(a) Test the manager's hypothesis.
(b) Construct a confidence interval for the population variance. Use $\alpha = 0.02$.

A: There are $N - 1 = 16$ degrees of freedom. For degrees of freedom less than 32 (small samples), the χ^2-distribution is used to test a null hypothesis or to construct a confidence interval for the variance of a normally distributed population (see Figure 1).

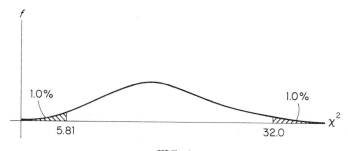

FIG. 1

(a) The null hypothesis is $H_0: \sigma^2 = 64$ and the alternative hypothesis is $H_a: \sigma^2 \neq 64$. For a 2 percent level of significance ($\alpha = 0.02$), two-tailed test, $df = N - 1 = 16$, the critical values are $\chi_{\alpha/2}^2 = 5.81$ and $\chi_{1-\alpha/2}^2 = 32$. The observed χ^2 is

$$\boxed{\chi^2 = \frac{(N-1)\,S^2}{\sigma_0^2}\,,}$$

where σ_0^2 is the hypothesized value. Then

$$\chi^2 = \frac{16 \cdot 100}{64} = 25.$$

The manager's hypothesis is not rejected.

(b) The confidence interval is

$$P\left(\chi_{\alpha/2}^2 < \frac{(N-1)\,S^2}{\sigma^2} < \chi_{1-\alpha/2}^2\right) = 1 - \alpha,$$

where $(N-1)\,S^2/\sigma^2$ has a χ^2-distribution with $df = N - 1 = 16$.

$$\boxed{P\left(\frac{(N-1)\,S^2}{\chi_{1-\alpha/2}^2} < \sigma^2 < \frac{(N-1)\,S^2}{\chi_{\alpha/2}^2}\right) = 1 - \alpha.}$$

$$P\left(\frac{16 \cdot 100}{32} < \sigma^2 < \frac{16 \cdot 100}{5.81}\right) = P(50 < \sigma^2 < 275) = 0.98.$$

There is a probability of 0.98 that the end points of the interval 50 to 275 bound the true population variance. Thus any null hypothesis about the size of the population variance that falls within this interval will be accepted.

Problem 2. Hypothesis Testing and Confidence Interval:
Variance, Large Sample

Q: Do the previous problem using a sample size of 73.

A: For large samples $(df > 32)$ the χ^2-distribution approaches the normal distribution, where the value of the standardized normal variable (Z) is found by the relation

$$Z = \sqrt{2\chi^2} - \sqrt{2df} - 1.$$

(a) The observed χ^2 is

$$\chi^2 = \frac{(N-1)S^2}{\sigma^2} = \frac{72 \cdot 100}{64} = 112.5.$$

Then

$$Z_0 = \sqrt{2\chi^2} - \sqrt{2df} - 1 = \sqrt{2 \cdot 112.5} - \sqrt{2 \cdot 72} - 1$$

$$\approx 15 - 12 = 3.$$

The acceptance interval is

$$P(Z_{\alpha/2} < Z_0 < Z_{1-\alpha/2}) = 1 - \alpha.$$

For $\alpha = 0.02$, $Z_{\alpha/2} = -2.33$ and $Z_{1-\alpha/2} = 2.33$. The null hypothesis is rejected.

Note: The same sample data result in the rejection of the null hypothesis $(H_0: \sigma^2 = 64)$ when the sample size is larger. Larger samples result in smaller acceptance intervals *ceteris paribus*. If a null hypothesis is rejected for a small sample, it will necessarily be rejected under a large sample, for the same sample values. If a null hypothesis is accepted for a large sample, it will necessarily be accepted for a small sample, given the same sample values.

(b) For large samples the confidence interval for σ^2 is

$$P\left(Z_{\alpha/2} < \sqrt{\frac{2(N-1)S^2}{\sigma^2}} - \sqrt{2df-1} < Z_{1-\alpha/2}\right) = 1 - \alpha.$$

Since $2(N-1) \approx 2df - 1$ for large samples,

$$P\left[Z_{\alpha/2} < \sqrt{2(N-1)}\left(\sqrt{\frac{S^2}{\sigma^2}} - 1\right) < Z_{1-\alpha/2}\right] = 1 - \alpha,$$

$$P\left(\frac{Z_{\alpha/2}}{\sqrt{2(N-1)}} + 1 < \frac{S}{\sigma} < \frac{Z_{1-\alpha/2}}{\sqrt{2(N-1)}} + 1\right) = 1 - \alpha,$$

$$P\left(\frac{S\sqrt{2df}}{Z_{1-\alpha/2} + \sqrt{2df}} < \sigma < \frac{S\sqrt{2df}}{Z_{\alpha/2} + \sqrt{2df}}\right) = 1 - \alpha.$$

Then for $df = 72$,

$$P\left(\frac{10\sqrt{2\cdot72}}{2.33 + \sqrt{2\cdot72}} < \sigma < \frac{10\sqrt{2\cdot72}}{-2.33 + \sqrt{2\cdot72}}\right) = 1 - \alpha,$$

and

$$P(8.37 < \sigma < 12.41) = 1 - \alpha,$$

$$P(70.06 < \sigma^2 < 154.01) = 0.98.$$

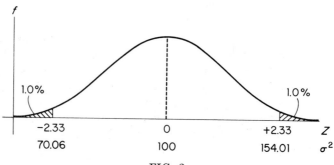

FIG. 2

Note: For the same sample values, larger samples result in smaller confidence intervals.

Problem 3. Hypothesis Testing: Equality of Two Variances

Q: A study of the distribution of education is concerned with the variance of years of schooling in New York and California. Given the following data from independent random samples, do the variances of years of schooling differ?

New York	California
$N_1 = 15$	$N_2 = 13$
$S_1^2 = 15$	$S_2^2 = 10$

Use a 10% level of significance ($\alpha = 0.10$).

A: A χ^2-distribution exists for each state. The ratio of two χ^2-distributions from independent samples divided by their respective degrees of freedom has a Fisher F-distribution.

$$H_0 : \sigma_1^2 = \sigma_2^2, \qquad H_a : \sigma_1^2 \neq \sigma_2^2$$

or

$$H_0 : \frac{\sigma_1^2}{\sigma_2^2} = 1, \qquad H_a : \frac{\sigma_1^2}{\sigma_2^2} \neq 1.$$

$F_{(\eta_1, \eta_2)}$ is the F-statistic, where η_1 designates the degrees of freedom in sample 1, η_2 designates the degree of freedom in sample 2.

$$F_{(\eta_1, \eta_2)} = \frac{(\chi_1^2 / \eta_1)}{(\chi_2^2 / \eta_2)} = \frac{(\eta_1 S_1^2 / \eta_1 \sigma_1^2)}{(\eta_2 S_2^2 / \eta_2 \sigma_2^2)} = \frac{(S_1^2 / \sigma_1^2)}{(S_2^2 / \sigma_2^2)}.$$

If the null hypothesis is true and there are no sampling fluctuations, $F = 1$.

The acceptance interval is

$$P(F_{\alpha/2} < F_0 < F_{1-\alpha/2}) = 1 - \alpha.$$

The null hypothesis is $\sigma_1^2 = \sigma_2^2$, so

$$F_0 = \left(\frac{S_1^2}{S_2^2}\right)\left(\frac{\sigma_2^2}{\sigma_1^2}\right) = \frac{S_1^2}{S_2^2} = \frac{15}{10} = 1.5.$$

The critical values are 0.40 and 2.64 (see Figure 3). From the table for the F-distribution (Table 3 in the Appendix),

$$F_{(14,12)}(0.95) = 2.64.$$

Tables for the F-distribution generally do not give the values for the lower end of the distribution. However

$$F_{(\eta_1,\eta_2)}(\alpha) = \frac{1}{F_{(\eta_2,\eta_1)}(1 - \alpha)}.$$

Thus

$$F_{(14,12)}(0.05) = \frac{1}{F_{(12,14)}(0.95)} = \frac{1}{2.53} = 0.40.$$

The null hypothesis of equal variances is accepted.

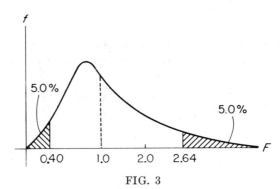

FIG. 3

Note: To avoid the computation of the critical F-value at the lower end of the distribution, a typical procedure is to place the larger sample variance in the numerator so that the observed F-ratio is greater than (or equal to) unity. The formula

$$P\left(F_{\alpha/2} < \left(\frac{S_1^2}{S_2^2}\right)\left(\frac{\sigma_2^2}{\sigma_2^2}\right) < F_{1-\alpha/2}\right) = 1 - \alpha$$

can be written as

$$\boxed{P\left(\frac{S_1^2}{S_2^2}\frac{\sigma_2^2}{\sigma_1^2} < F_{1-\alpha/2}\right) = 1 - \alpha}$$

since $F_{\alpha/2}$ must be less than unity. Then, for this problem,

$$P\left[\left(\frac{S_1^2}{S_2^2}\right)\left(\frac{\sigma_2^2}{\sigma_1^2}\right) < 2.64\right] = 0.90,$$

and since $(S_1^2/S_2^2)(\sigma_2^2/\sigma_1^2) = 1.5$, the null hypothesis is accepted.

Problem 4. Confidence Interval: Ratio of Two Variances

Q: It has been hypothesized that the variance of schooling is k times larger in the South than in the non-South. Construct a 98 percent confidence interval for k given the data on two independent random samples:

Non-South	South
$N_1 = 20$	$N_2 = 16$
$S_1^2 = 15$	$S_2^2 = 10$

A: $P\left(F_{(\eta_1,\eta_2)}(\alpha/2) < \left(\frac{S_1^2}{S_2^2}\right)\left(\frac{\sigma_2^2}{\sigma_1^2}\right) < F_{(\eta_1,\eta_2)}(1 - \alpha/2)\right) = 1 - \alpha,$

$k = \dfrac{\sigma_2^2}{\sigma_1^2}$ and $\eta_1 = 19, \eta_2 = 15,$

$F_{(\eta_1,\eta_2)}(0.99) = 3.36$

and

$$F_{(\eta_1, \eta_2)}(0.01) = \frac{1}{F_{(\eta_2, \eta_1)}(0.99)} = \frac{1}{3.15} = 0.32,$$

$$P\left(F(\alpha/2)\,\frac{S_2^2}{S_1^2} < k < F(1-\alpha/2)\,\frac{S_2^2}{S_1^2}\right) = 0.98,$$

$$P\left(0.32(\tfrac{10}{15}) < k < 3.36(\tfrac{10}{15})\right) = 0.98,$$

$$P(0.21 < k < 2.24) = 0.98.$$

The probability is 0.98 that the true value of k is bounded by the values 0.21 and 2.24.

Problem 5. Hypothesis Testing: Equality of Several Variances, Equal Sample Size

Q: A random survey of monthly food expenditures (X) in four census areas revealed the following data

Area	$df = N - 1$	S_x^2
1	30	100
2	30	90
3	30	60
4	30	40

Does the variance in expenditures differ *across* the four $(k = 4)$ areas?

A: $H_0: \sigma_1^2 = \sigma_2^2 = \sigma_3^2 = \sigma_4^2$. The F-ratio cannot be used to test the equality of variances since there are more than two samples. A special distribution, $F_{\max}$, has been developed for this purpose when the sample sizes are the same. The critical value for the $F_{\max}$ distribution is a function of the number of degrees of freedom in one sample (df) and the number of samples (k).

$$\text{Observed } F_{\max} = \frac{S_{\max}^2}{S_{\min}^2} = \frac{100}{40} = 2.5.$$

For a 5 percent level of significance, $df = 30$ and $k = 4$, critical $F_{max} = 2.61$. (See Appendix Table 6.) The observed F_{max} is less than the critical F_{max}. The null hypothesis of equal variances is not rejected.

Problem 6. Hypothesis Testing: Equality of Several Variances, Unequal Sample Size

Q: Do Problem 5 using the following data:

Area	$df = N - 1$	S_x^2
1	30	100
2	50	90
3	60	60
4	15	40

A: $H_0: \sigma_1^2 = \sigma_2^2 = \sigma_3^2 = \sigma_4^2$. When samples sizes differ, the F_{max} test can be applied, but does not always provide a solution. The observed F_{max} is $F_{max} = \frac{100}{40} = 2.5$.

For a 5 percent level of significance and the largest number of degrees of freedom $(df = 60)$, $F_{max}(60,4) = 1.96$. For the largest number of degrees of freedom the null hypothesis is rejected. If the null hypothesis were accepted in this situation, it would certainly be accepted for a smaller number of degrees of freedom.

For the smallest number of degrees of freedom $(df = 15)$, $F_{max}(15,4) = 4.01$, which exceeds the observed value. The null hypothesis is accepted. If, however, the null hypothesis were not accepted for the smallest sample, it would not be accepted for any sample.

In the current example the hypothesis of equal variances is accepted for the smallest sample but rejected for the largest sample, so the F_{max} test gives ambiguous results. This situation can be treated by using Bartlett's test. (See footnote on p. 111.)

Problem 7. Testing Goodness of Fit to a Hypothesized Distribution

Q: It has been suggested by a vice-president of a large corporation that the marital status of the executives is 30 percent married, 20 percent never married, 30 percent divorced, and 20 percent widowed. A statistician reports that a random sample of 200 indicates 26 percent married, 24 percent never married, 25 percent divorced, and 25 percent widowed.

(a) Test the vice-president's hypothesis.
(b) Test the hypothesis that the statistician gave the firm a fake uniform distribution rather than a sample randomly selected from a uniform distribution.

A: Both parts are a test of the goodness of fit of an observed frequency distribution to a theoretical distribution. If k is the number of types of marital status, the tests involve constructing a χ^2 with $k - 1$ degrees of freedom.

(a) Let M = married, NM = never married, D = divorced, and W = widowed. Our frequencies are:

	M	NM	D	W
Observed (f_i)	52	48	50	50
Expected (theoretical) (F_i)	60	40	60	40

$$\chi_{k-1}^2 = \sum_{i=1}^{k} \frac{(f_i - F_i)^2}{F_i}.$$

Observed $\chi_{4-1}^2 = \dfrac{8^2}{60} + \dfrac{8^2}{40} + \dfrac{10^2}{60} + \dfrac{10^2}{40} = \dfrac{820}{120} = 6.8.$

For a type I error of 5 percent, the critical $\chi^2(0.95) = 7.8$, as illustrated in Figure 4. The vice-president's hypothesis is not rejected.

This is a *one*-tailed test, with only a right tail. The null hypothesis is rejected only if the deviations of the observed distribution from the expected distribution are "too large,"

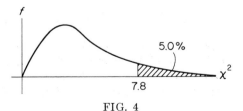

FIG. 4

that is, if the probability of obtaining by random sampling the observed χ^2-value (or a larger value) is very small.

(b) Here we have the following frequencies:

	M	NM	D	W
Observed (f_i)	52	48	50	50
Expected (F_i)	50	50	50	50

The object of part (b) is to see whether the fit of the observed distribution to the hypothesized distribution (in this case the uniform distribution) is so close that the sample is not likely to be the result of random sampling from the hypothesized distribution.

$$\chi_{k-1}^2 = \sum_{i=1}^{k} \frac{(f_i - F_i)^2}{F_i} = \frac{2^2}{50} + \frac{2^2}{50} + \frac{0^2}{50} + \frac{0^2}{50}.$$

Observed χ^2: $\chi_3^2 = \frac{8}{50} = 0.16$,
   critical χ^2: $\chi_3^2(0.05) = 0.35$ and $\chi_3^2(0.01) = 0.11$.

The test involves one tail—the *left* tail. Is the probability of obtaining the observed frequencies from random sampling of the hypothesized population so small that it is not likely that random sampling did in fact occur? Random sampling from a uniform distribution ($k = 4$) would give χ^2-values of 0.11 or less 1 percent of the time, and χ^2-values of 0.35 or less 5 percent of the time. The observed χ^2 is 0.16. Thus, if a 5 percent level of significance is used, the hypothesis is accepted that the data are "too good" a fit to the uniform distribution for the data to be the result of random sampling (see Figure 5).

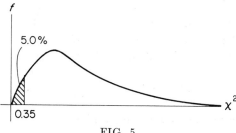

FIG. 5

Problem 8. Testing whether Several Samples Come from the Same
Population (Homogeneous χ^2)

Q: Samples of household income are taken in four cities. Do the samples
come from the same population?

City

	I	II	III	IV	Total
Under $3,000	10	15	15	10	50
$3,000–$8,000	5	10	15	10	40
Over $8,000	15	15	10	20	60
Total	30	40	40	40	150

A: A test of whether the samples come from the same population is
often called a *test of homogeneity* or a *test of independence*. The pooled
sample is used as an estimate of the population distribution.

Our expected frequencies (sample sizes multiplied by expected
relative frequencies) are:

	I	II	III	IV	Total
Under $3,000	$30\left(\frac{5}{15}\right)$	$40\left(\frac{5}{15}\right)$	$40\left(\frac{5}{15}\right)$	$40\left(\frac{5}{15}\right)$	50
$3,000–$8,000	$30\left(\frac{4}{15}\right)$	$40\left(\frac{4}{15}\right)$	$40\left(\frac{4}{15}\right)$	$40\left(\frac{4}{15}\right)$	40
Over $8,000	$30\left(\frac{6}{15}\right)$	$40\left(\frac{6}{15}\right)$	$40\left(\frac{6}{15}\right)$	$40\left(\frac{6}{15}\right)$	60
Total	30	40	40	40	150

Multiplying, we get

	I	II	III	IV	Total
Under $3,000	10	13.33	13.33	13.33	50
$3,000–$8,000	8	10.67	10.67	10.67	40
Over $8,000	12	16.00	16.00	16.00	60
Total	30	40.00	40.00	40.00	150

If we let r = number of rows (intervals) and c = number of columns (samples), then

$$\chi_H{}^2 = \sum_{i=1}^{rc} \frac{(f_i - F_i)^2}{F_i}$$

$$= \frac{(10 - 10)^2}{10} + \frac{(5 - 8)^2}{8} + \frac{(15 - 12)^2}{12}$$

$$+ \frac{(15 - 13.33)^2}{13.33} + \frac{(10 - 10.67)^2}{10.67} + \frac{(15 - 16)^2}{16}$$

$$+ \frac{(15 - 13.33)^2}{13.33} + \frac{(15 - 10.67)^2}{10.67} + \frac{(10 - 16)^2}{16}$$

$$+ \frac{(10 - 13.33)^2}{13.33} + \frac{(10 - 10.67)^2}{10.67} + \frac{(20 - 16)^2}{16}$$

$$= 0 + \frac{3^2}{8} + \frac{3^2}{12} + \frac{(1.67)^2}{13.33} + \frac{(0.67)^2}{10.67} + \frac{1^2}{16} + \frac{(1.67)^2}{13.33}$$

$$+ \frac{(4.33)^2}{10.67} + \frac{6^2}{16} + \frac{(3.33)^2}{13.33} + \frac{(0.67)^2}{10.67} + \frac{4^2}{16}$$

$$= 8.3.$$

Observed Homogeneous χ^2: $\chi^2 = 8.3$.

We reject the null hypothesis that the samples come from the same population if the observed homogeneous $\chi_H{}^2$ exceeds the critical χ^2 (see Figure 6).

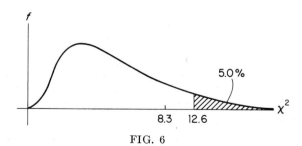

FIG. 6

There are $(r - 1)(c - 1)$ degrees of freedom, so

$$df = (3 - 1)(4 - 1) = 6.$$

Once the expected frequencies are computed from the pooled data, one of the samples is not independent.

Critical χ^2: $\chi_6^2(0.95) = 12.6$.

The hypothesis that the samples come from the same population cannot be rejected.

Problem 9. Testing Goodness of Fit of Several Large Samples to a
               Hypothesized Distribution (Combined χ^2)

Q: Using the data given in the previous problem, test the hypothesis that income is uniformly distributed among the three income groups.

A: The χ^2 used to test the hypothesis that several samples are from the same population with a particular distribution is called a *combined* χ^2 and is designated χ_C^2.

  For the sample of 150, the expected frequencies (sample sizes times expected relative frequencies or proportions) are:

	I	II	III	IV	Total	
Under $3,000	$30\,(\frac{1}{3})$	$40\,(\frac{1}{3})$	$40\,(\frac{1}{3})$	$40\,(\frac{1}{3})$	50	
$3,000–$8,000	$30\,(\frac{1}{3})$	$40\,(\frac{1}{3})$	$40\,(\frac{1}{3})$	$40\,(\frac{1}{3})$	50	Hypothesized
Over $8,000	$30\,(\frac{1}{3})$	$40\,(\frac{1}{3})$	$40\,(\frac{1}{3})$	$40\,(\frac{1}{3})$	50	values
Total	30	40	40	40	150	

Or:

	I	II	III	IV	Total
Under $3,000	10	13.33	13.33	13.33	50
$3,000–$8,000	10	13.33	13.33	13.33	50
Over $8,000	10	13.33	13.33	13.33	50
Total	30	40.00	40.00	40.00	150

The observed combined χ_c^2 $\left(\chi_c^2 = \sum_{i=1}^{rc} [(f_i - F_i)^2/F_i]\right)$, where f_i is the observed frequency and F_i is the expected frequency, is compared with the critical χ^2. There are $df = (r-1)c = (3-1)(4) = 8$ degrees of freedom, since one degree of freedom is lost in each sample. (r = number of intervals, c = number of samples.)

$$\chi_c^2 = \sum_{i=1}^{rc} \frac{(f_i - F_i)^2}{F_i}$$

$$= \frac{(10 - 10)^2}{10} + \frac{(5 - 10)^2}{10} + \frac{(15 - 10)^2}{10}$$

$$+ \frac{(15 - 13.33)^2}{13.33} + \frac{(10 - 13.33)^2}{13.33} + \frac{(15 - 13.33)^2}{13.33}$$

$$+ \frac{(15 - 13.33)^2}{13.33} + \frac{(15 - 13.33)^2}{13.33} + \frac{(10 - 13.33)^2}{13.33}$$

$$+ \frac{(10 - 13.33)^2}{13.33} + \frac{(10 - 13.33)^2}{13.33} + \frac{(20 - 13.33)^2}{13.33}$$

$$= \frac{2(5)^2}{10} + \frac{4(1.67)^2 + 4(3.33)^2 + (6.67)^2}{13.33}$$

$$= 12.5.$$

Observed combined χ^2: $\chi_c^2 = 12.5$,
         critical χ^2: $\chi_8^2(0.95) = 15.5$.

The hypothesis that the samples come from the same population which has a uniform distribution is accepted (see Figure 7).

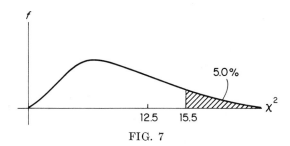

FIG. 7

Problem 10. Testing Goodness of Fit of Pooled Data to a
Hypothesized Distribution (Pooled χ^2)

Q: In the preceding problem we accepted the null hypothesis that the
four samples came from a single population with a particular, in
this case a uniform, distribution. Pool the sample data and test the
hypothesis that the data are consistent with a uniform frequency
distribution.

A: Our frequencies are:

	Observed	Expected
Under $3,000	50	50
$3,000–$8,000	40	50
Over $8,000	60	50
Total	150	150

The observed pooled χ^2 is

$$\chi_P^2 = \sum_{i=1}^{r} \frac{(f_i - F_i)^2}{F_i} = \frac{0}{50} + \frac{10^2}{50} + \frac{10^2}{50} = \frac{200}{50} = 4.$$

There are $df = r - 1 = 2$ degrees of freedom, where r is the number
of intervals. The number of degrees of freedom is one less than the
number of intervals in the distribution. The critical values are
$\chi_2^2(0.95) = 2.92$ and $\chi_2^2(0.975) = 4.30$.

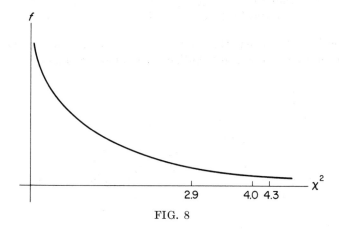

FIG. 8

The hypothesis of a uniform distribution would be rejected under a 5 percent level but not under a 2.5 percent level of significance (see Figure 8).

Note: The χ^2-values and degrees of freedom for the combined, pooled, and homogeneous χ^2 are interrelated:

$$
\begin{array}{lll}
\text{Pooled} + \text{Homogeneous} & = & \text{Combined} \\
\chi_P^2 \quad + \chi_H^2 & = & \chi_C^2 \\
df_P \quad + df_H & = & df_C
\end{array}
$$

These three values can be summarized as follows: Combined χ^2 tests whether several samples are from the same population with a particular distribution. Pooled χ^2 tests whether the samples come from a particular distribution, given that the data are from the same population. Homogeneous χ^2 tests whether the samples are from the same population without any hypothesis about the distribution in the population. Note also that absolute frequencies not relative frequencies are used in computing the value of χ^2. In the example used here:

$$\chi_H^2 + \chi_P^2 = 8.3 + 4.0 = 12.3, \qquad \chi_C^2 = 12.5.$$

(The difference between 12.3 and 12.5 is due to rounding errors!) And

$$df_H + df_P = 6 + 2 = 8, \qquad df_C = 8.$$

Problem 11. Testing Goodness of Fit of Two or More Small
              Samples to a Hypothesized Distribution

Q: In a sample of six pensioners, four supported a large increase in social security benefits. In a sample of nine young workers, only one supported the increase. Can you accept the hypothesis that both pensioners and young workers are equally divided in their support for the increase?

A: The observed frequencies are:

	Favor	Oppose	Total
Pensioners	4	2	6
Young workers	1	8	9
Total	5	10	15

The null hypothesis is that $P = \frac{1}{2}$, that is, half the population of pensioners and half that of young workers support the increase.

When the number of observations in many cells is small (less than 5) the calculated χ^2 is a poor approximation to the theoretical χ^2-distribution. There are two solutions, the Yates Correction and the Exact Method.

(a) The *Yates Correction* can be used only when we have two rows and two columns in our frequency table. Such a table is called a *two-by-two contingency table*. We keep the marginals the same but decrease the observed χ^2 by adding or subtracting $\frac{1}{2}$ to each cell:

Corrected observed

Favor	Oppose	Total
3.5	2.5	6
1.5	7.5	9
5.0	10.0	15

Expected

Favor	Oppose	Total
$6(\frac{1}{2})$	$6(\frac{1}{2})$	6
$9(\frac{1}{2})$	$9(\frac{1}{2})$	9
7.5	7.5	15

Then

$$\text{observed } \chi^2 = \sum_{i=1}^{4} \frac{(f_i - F_i)^2}{F_i}$$

$$= \frac{(3.5 - 3)^2}{3} + \frac{(2.5 - 3)^2}{3}$$

$$+ \frac{(1.5 - 4.5)^2}{4.5} + \frac{(7.5 - 4.5)^2}{4.5}$$

$$= 4.17.$$

For critical χ^2 ($df = 1$),

$$\chi_1^2(0.95) = 3.8, \qquad \chi_1^2(0.975) = 5.0.$$

The null hypothesis is rejected under a 5 percent level but not under a 2.5 percent level of significance.

(b) The *Exact Method* can be used for *any* size contingency table. Keeping marginal values fixed, find the possible cell combinations that are "worse" than the observed values in terms of the null hypothesis. The sum of the probabilities of "worse" possibilities is the type I (alpha) error. If the sum of these probabilities is small, we reject the null hypothesis. The "worse" situations are:

$$
\text{I:} \quad \begin{array}{c|c} 5 & 1 \\ \hline 0 & 9 \end{array} \qquad \text{II:} \quad \begin{array}{c|c} 4 & 2 \\ \hline 1 & 8 \end{array}
$$

$P(\text{I})$ or $P(\text{II})$

$$= \frac{P(\text{outcome for pensioners})\,P(\text{outcome for young workers})}{P(\text{observed total outcome})}$$

$$P(\text{I}) = \frac{\left[\binom{6}{5} P^5 Q^1\right]\left[\binom{9}{0}(P)^0 Q^9\right]}{\binom{15}{5} P^5 Q^{10}},$$

$$P(\text{II}) = \frac{\left[\binom{6}{4} P^4 Q^2\right]\left[\binom{9}{1} P^1 Q^8\right]}{\binom{15}{5} P^5 Q^{10}},$$

where P and Q are the hypothesized values, $P = Q = \frac{1}{2}$. Adding, we get

$$P(\text{I}) + P(\text{II}) = 0.047.$$

The probability of obtaining a "worse" result is very small (4.7 percent). The null hypothesis is rejected under a 5 percent level of significance.

Note: The table of factorials (Table 10 of the Appendix) facilitates the computation of probabilities.

Part B

Regression Analysis

Chapters 3 and 4 presented procedures for testing hypotheses about the relationships between the means and variances of two or more variables. A common objective of statistical analysis is to test for relationships among the values of two or more variables. For example, an investigator might be interested in whether height, weight, and eye color are associated with each other. The analysis of the relationships among the data on two or more variables is referred to as *multivariate statistical analysis*. The most commonly used technique for multivariate analysis in economics and business, and increasingly in the other social sciences, is *linear regression analysis*. Regression analysis is used for testing hypotheses about the relationship between two variables and for prediction. Part B develops the procedures for using this technique.

In a regression model one variable (Y_i) is written as a linear function of one or more other variables (X_i) and an unexplained residual (U_i):

$$Y_i = \beta_0 + \beta_1 X_i + U_i.$$

The variable Y_i is called the *dependent variable* because it is expressed as a function of another variable, called the *independent* or *explanatory variable*. If there is a functional relationship between two variables they are said to be *associated*. Regression analysis tests for *linear association*, which is also called *correlation*.

In statistical analysis dependence is a purely mathematical concept and should not be confused with causation. *Causation* implies a behavioral relationship between two variables. If a linear regression expresses Y as "dependent on" X, and if it is found empirically that there is an association between Y and X, we still do not know whether X caused Y, Y caused X, or another variable, Z, caused X and Y. Inferences as to causal relationships between the variables must be based on information from other sources.

Regression analysis permits the division of the variation in the dependent variable (Y) into a component that can be attributed to the explanatory variables $(\hat{Y})$ and a component that cannot be attributed to the explanatory variables (U):

$$Y = \hat{Y} + U.$$

The regression model's "explanatory" power (R^2) is the ratio of the variation in the dependent variable attributable to the explanatory variables divided by the total variation in the dependent variable:

$$R^2 = \frac{S^2(\hat{Y})}{S^2(Y)} .$$

This explanatory power refers to the extent of a statistical linear association, and should not be interpreted as a measure of the extent of causation.

A *simple regression* contains one explanatory variable, and a *multiple regression* contains two or more explanatory variables. The simple regression model is seldom used in actual statistical analysis because it is rare that an investigator wishes to include only one explanatory variable. The basic principles of regression analysis are the same for a simple and a multiple regression. It is, however, easier to learn the techniques and principles of regression analysis by first focusing on the simple form. Once this is mastered, the additional techniques and principles associated with multiple regression analysis are quickly learned.

The simple regression model is developed in Chapter 5. Chapter 6 presents the analysis for multiple regression. Under certain circumstances a single equation is inappropriate and the empirical analysis must be performed for a set of interrelated equations (*simultaneous system*). This is the subject of Chapter 7.

The presentation of the regression analysis in Part B assumes that the reader is familiar with the basic principles of probability and hypothesis testing developed in Part A, especially in Chapter 3. It would be foolhardy to attempt to learn regression analysis without first understanding these basic building blocks.

Chapter 5

Simple Regression Analysis

In the simple regression model the dependent variable Y is assumed to be a linear function of only one explanatory variable, X_i. The regression model is written as

$$Y_i = \beta_0 + \beta_1 X_i + U_i,$$

where β_0 and β_1 are the unknown population coefficients, the subscript i designates the ith observation in the population, and U_i is the residual or unexplained part of Y_i. The estimated regression equation is written as

$$Y_i = b_0 + b_1 X_i + \hat{U}_i,$$

where b_0 and b_1 are the sample estimates of the population coefficients β_0 and β_1, the subscript i designates the ith observation in the sample, and $\hat{U}_i$ is the estimated residual for the ith observation. The computed values for the sample regression coefficients are used as estimates of the true, but unknown, values of the population parameters. The object of regression analysis is to compute the sample coefficients b_0 and b_1 which "best" fit the data in the sense that they maximize the explanatory power of the regression model.

The maximization of the regression model's explanatory power is the same as the minimization of the sum of the squared residuals $(\sum_{i=1}^{N} \hat{U}_i^2)$ for a given set of data. This is why regression analysis is often referred to as *least-squares analysis*.

Rewriting the sample regression equation, we have

$$\hat{U}_i = Y_i - b_0 - b_1 X_i.$$

Squaring both sides and summing across all of the N observations in the sample gives

$$\sum_{i=1}^{N} \hat{U}_i^2 = \sum_{i=1}^{N} (Y_i - b_0 - b_1 X_i)^2.$$

Taking the first derivative of the residual sum of squares with respect to b_0 and b_1, we get

$$\frac{\partial \sum_{i=1}^{N} \hat{U}^2}{\partial b_0} = 2 \sum_{i=1}^{N} (Y_i - b_0 - b_1 X_i)(-1),$$

$$\frac{\partial \sum_{i=1}^{N} \hat{U}^2}{\partial b_1} = 2 \sum_{i=1}^{N} (Y_i - b_0 - b_1 X_i)(-X_i).$$

Setting the first derivatives equal to zero and rearranging terms then yields

$$\boxed{\bar{Y} = b_0 + b_1 \bar{X}}$$

and

$$\boxed{\sum_{i=1}^{N} X_i Y_i = b_0 \sum_{i=1}^{N} X_i + b_1 \sum_{i=1}^{N} X_i^2.}$$

These two equations contain two unknowns, b_0 and b_1, and can be solved for unique solutions for the estimates of the regression coefficients. The equations are called the *normal equations* of the regression model.

The regression equation is frequently written with the variables expressed as deviations from their means. When lower-case letters are used to designate deviations from means—$x_i = X_i - \bar{X}$, $y_i = Y_i - \bar{Y}$, and $\hat{u}_i = \hat{U}_i - \bar{\hat{U}}$†—the regression equation is

$$y_i = b_1 x_i + \hat{u}_i,$$

and the two normal equations are

$$b_1 = \frac{\displaystyle\sum_{i=1}^{N} x_i y_i}{\displaystyle\sum_{i=1}^{N} x_i^2}, \qquad \bar{Y} = b_0 + b_1\bar{X}.$$

In the least-squares linear regression model there are four assumptions about the distribution of the residual in the population (see Figure 1). They are:

1. The residual is normally distributed with a zero mean.
2. The residual is uncorrelated with the explanatory variable.
3. The residual variance $[S^2(U)]$ is constant for all values of X.
4. The values of the residual are not correlated with each other.

What would happen if each of these assumptions did not hold?

Assumption 1. If the population residual is not normally distributed and if the sample size is small, t-tests and F tests for the significance of the regression coefficients are not strictly applicable. For example, the estimated slope coefficient b_1 is a value drawn from a sampling distribution. Since

$$b_1 = \frac{\displaystyle\sum_{i=1}^{N} x_i y_i}{\displaystyle\sum_{i=1}^{N} x_i^2},$$

† Since $\bar{\hat{U}} = 0$ in the regression model, $\hat{u}_i = \hat{U}_i$.

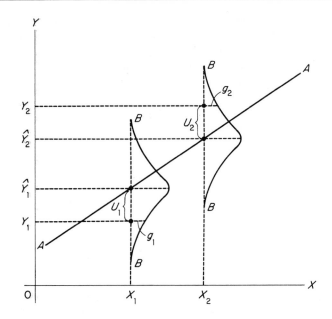

AA: Regression line in the population,

$$Y_i = \beta_0 + \beta_1 X_i + U_i = \hat{Y}_i + U_i$$

BB: Normal distribution of the residual around the regression line

g_i : Frequency of the residual for observation i

FIG. 1. The regression line in the population.

viewing the x_i's as fixed, b_1 is a *linear* function of the dependent variable Y_i and hence of the residual $\hat{U}_i$. As indicated in Chapter 3, if the residuals are normally distributed or if the sample size is large, due to random sampling the linear combination b_1 has a sampling distribution that is approximately normal.†

Assumption 2. If the residual in the population is correlated with the explanatory variable, the regression coefficient is a biased estimate of

† The regression procedure is applicable only to randomly selected samples.

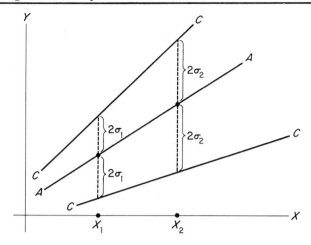

σ_i^2 = Residual variance for the ith value of X

AA: Regression line in the population

CC: Points two standard deviations of the residual from the regression line

FIG. 2. Heteroscedastic residuals.

the true population parameter.† If the estimate is *unbiased*, the estimate is just as likely to be greater than the true value as it is to be less than the true value, and the mean of a very large number of estimates obtained from random sampling would be approximately equal to the true population parameter.

† Using the equation for the estimate of the slope coefficient, replacing $\overset{N}{\underset{i=1}{\Sigma}}$ by Σ, we have

$$b_1 = \frac{\Sigma\, x_i y_i}{\Sigma\, x_i^2} = \frac{\Sigma\, x_i(\beta x_i + \hat{u}_i)}{\Sigma\, x_i^2} = \beta_1 + \frac{\Sigma\, x_i(\hat{u}_i)}{\Sigma\, x_i^2}.$$

If $w_i = x_i/\Sigma\, x_i^2$, then $b_1 = \beta_1 + \Sigma\, w_i \hat{u}_i$. Taking expected values gives $E(b_1) = E(\beta_1 + \Sigma\, w_i \hat{u}_i) = \beta_1 + E(\Sigma\, w_i \hat{u}_i)$. The expected value of the estimated slope coefficient equals the population value if in the population the residual and the explanatory variable are uncorrelated. The expected value of a variable is the mean value of the variable obtained from a very large number of samples of a given size.

The residual and the explanatory variable are correlated only if there is a variable that is correlated with both the explanatory variable and the dependent variable. The bias is eliminated if this variable is explicitly included in the regression equation. This involves a multiple regression (see Chapter 6, especially Problem 1).

Assumption 3. When the residual variance depends on the value of the explanatory variable X, the residual is said to be *heteroscedastic* (see Figure 2). When the residual variance is independent of the value of the explanatory variable, the residual is said to be *homoscedastic*. Heteroscedastic residuals do not bias the estimates of the regression coefficients, but they do bias the estimates of the standard errors of the regression coefficients. This results in biased tests of hypotheses about the regression coefficients and biased estimates of the reliability of predictions based on the estimated equation. *Weighted regressions* (generalized least-squares) can be used to compute a least-squares regression with homoscedastic residuals (see Problems 6 and 7 of this chapter).

Assumption 4. When the residuals in the population are correlated with each other, they are said to be *autocorrelated*. Autocorrelated residuals are common in time-series analysis, where they are referred to as *serially correlated residuals* (see Figure 3). Like heteroscedastic residuals, autocorrelated residuals do not bias the regression coefficients, but do bias the standard errors of the regression coefficients. There are several

A dot designates an observation
of Y for the time period X_i

FIG. 3. Positive serial correlation in the residuals.

procedures that can be used to eliminate autocorrelation in the residuals (see Problem 11 of Chapter 6).

When all four assumptions of the least-squares regression model are satisfied, the sample estimates of the population parameters have certain desirable properties. The sample estimates are BLUE, that is, they are Best Linear Unbiased Estimates of the true population parameters.†️ Linear and unbiased estimates have already been defined (see p. 140–141). "Best" means that (for the class of linear unbiased estimators) the procedure results in an estimate of the regression coefficient with the smallest sampling variability, and hence the greatest reliability.

The relatively low cost of computers and the availability of easily used pre-programmed statistical packages have made hand calculation—by pencil and paper or desk calculator—of regression equations an anachronism. For learning the principles of regression analysis, however, there is no substitute for actually computing each part by "hand." To make this task easier, this chapter is devoted to the development of the simple regression model and its application to a particular problem with a very small sample in which the data arecomputationally simple. The reader is encouraged to take advantage of this opportunity to do each of the calculations by "hand," and then, if he or she wishes, to use a computer program on the same set of data.

Data. In a study of fertility patterns a random sample of ten newly married couples were asked the number of children they desired to have (X). Twenty years later all ten couples were asked the number of children they actually had (Y). The results are given in Table 1. It is "known" that the residuals in the population are normally distributed.‡ The investigator is interested in examining whether the number of children each couple eventually had (Y) can be explained by the number they said they desired at the time of their marriage.

---

† For a rigorous proof of these properties, see any standard econometrics textbook with a theoretical orientation. See, for example, Teh-wei Hu, *Econometrics: An Introductory Analysis* (University Park Press, Baltimore, 1973) pp. 42–44.

‡ Regression equations are generally not computed for very small samples. When small samples are used, however, the observed residuals can be tested to see whether they are consistent with sampling from a normal distribution. If the residuals in the small sample case are not consistent with a normal distribution of residuals in the population, the conventional t-tests and F-tests of regression analysis are not strictly applicable.

TABLE 1
Actual and Desired Number of Children of Ten Rendomly Selected Couples

	Number of children	
Couple	Actual (Y)	Desired (X)
1	0	0
2	2	1
3	1	2
4	3	1
5	1	0
6	3	3
7	4	4
8	2	2
9	1	2
10	2	1

Problems

1. Simple Regression Equation
2. Hypothesis Testing of Regression Coefficients
3. Coefficient of Determination
4. Coefficient of Correlation: Hypothesis Testing and Confidence Interval
5. Reversing the Order of the Regression Equation
6. Heteroscedastic Residuals
7. Prediction from a Simple Regression
8. Functional Form
9. Errors in Variables

Problem 1. Simple Regression Equation

Q: (a) Compute the linear regression of the actual number of children each couple had (Y) on the number they said they desired to have at the time of marriage (X).

(b) Test whether the regression coefficients differ significantly from zero at a 5 percent level of significance using the t-test.

(c) Show that the t-test for the significance of the slope coefficient in a simple regression is related to the F-distribution.

(d) What is the elasticity of the dependent variable with respect to the explanatory variable at the mean?

(e) In a summary table give the regression coefficients, standard errors, and t-ratios for tests of significance.

A: (a) The population equation is written as

$$Y_i = \beta_0 + \beta_1 X_i + U_i.$$

The sample regression equation is written as

$$Y_i = b_0 + b_1 X_i + \hat{U}_i \quad \text{or} \quad \hat{Y}_i = b_0 + b_1 X_i.$$

In terms of deviations from means, if $x_i = X_i - \bar{X}$ and $y_i = Y_i - \bar{Y}$, the normal equations are

$$\sum_{i=1}^{N} x_i y_i = b_1 \sum_{i=1}^{N} x_i^2 \quad \text{and} \quad \bar{Y} = b_0 + b_1 \bar{X}.$$

The sums of variables needed to estimate the regression parameters are presented in Table 2. Substituting the appropriate

TABLE 2

Computation of the Data Used in the Regression Analysis

Couple	X_i	$x_i =$ $(X_i - \bar{X})$	$x_i^2 =$ $(X_i - \bar{X})^2$	Y_i	$y_i =$ $(Y_i - \bar{Y})$	$x_i y_i =$ $(X_i - \bar{X})(Y_i - \bar{Y})$	$y_i^2 =$ $(Y_i - \bar{Y})^2$
1	0	-1.6	2.56	0	-1.9	3.04	3.61
2	1	-0.6	0.36	2	0.1	-0.06	0.01
3	2	0.4	0.16	1	-0.9	-0.36	0.81
4	1	-0.6	0.36	3	1.1	-0.66	1.21
5	0	-1.6	2.56	1	-0.9	1.44	0.81
6	3	1.4	1.96	3	1.1	1.54	1.21
7	4	2.4	5.76	4	2.1	5.04	4.41
8	2	0.4	0.16	2	0.1	0.04	0.01
9	2	0.4	0.16	1	-0.9	-0.36	0.81
10	1	-0.6	0.36	2	0.1	-0.06	0.01
$N = 10$	16	0.0	14.40	19	0.0	+9.60	12.90

$$\bar{X} = \frac{\sum_{i=1}^{N} X_i}{N} = 1.6, \quad \bar{Y} = \frac{\sum_{i=1}^{N} Y_i}{N} = 1.9, \quad \sum_{i=1}^{N} x_i^2 = 14.40, \quad \sum_{i=1}^{N} x_i y_i = 9.60.$$

values into the normal equations yields

$$b_1 = \frac{\sum\limits_{i=1}^{N} x_i y_i}{\sum\limits_{i=1}^{N} x_i^2} = \frac{9.60}{14.40} = 0.667,$$

$$b_0 = \bar{Y} - b_1 \bar{X} = 1.90 - (0.667)(1.60) = 0.833.$$

The regression equation is

$$\hat{Y}_i = 0.833 + 0.667 X_i.$$

The data points and the regression line are shown in Figure 4·

(b) To test the statistical significance of a regression coefficient is to test the null hypothesis that the population value is zero. If the deviation of the regression coefficient (b_1) from its hypothesized value (β_1) is small relative to its sampling variability $[S(b_1)]$, the null hypothesis cannot be rejected. The statistical significance of the slope and intercept of a simple regression equation is tested using the t-distribution, with $N - 2$ degrees of freedom; one degree of freedom is lost for each estimated regression coefficient.

Recall that $t = (b_1 - \beta_1)/S(b_1)$ has a t-distribution if the regression coefficient (b_1) is normally distributed. The regression coefficients are linear combinations of the dependent variable Y. Thus for large samples $(N > 30)$ the sampling distribution of a regression coefficient is normally distributed regardless of the distribution of the residual. For small samples $(N < 30)$ the sampling distribution of the regression coefficient is normally distributed only if the dependent variable Y for given values of X is normally distributed, that is, only if the residual $\hat{U}$ is normally distributed. If the hypothesis that the residuals are from a normal distribution is not rejected, the assumption of normality can be made (see test of goodness of fit, Chapter 4).

Although in this problem the sample size is small, we have been told that the residuals in the population are normally distributed. Thus the sampling distribution of the regression coefficients is normal, and the t-distribution may be used to test hypotheses about the values of the population parameters.

For a simple regression the square of the standard error of the slope coefficient† is

$$
S^2(b_1) = \frac{S^2(\hat{U})}{\displaystyle\sum_{i=1}^{N} x_i^2},
$$

† If the symbol E is used to designate an expected value (i.e., a mean value in a population) and if the residuals are homoscedastic $[E(\hat{u})^2 = \sigma^2$ for all values of $X]$, not autocorrelated $[E(\hat{u}_i, \hat{u}_j) = 0]$, and uncorrelated with the explanatory variable in the population $[E(x_i, \hat{u}_i) = 0]$, then

$$
S^2(b_1) = E[(b_1 - \beta_1)^2] = E\left[\frac{(\Sigma x_i \hat{u}_i)^2}{(\Sigma x_i^2)^2}\right]
$$

$$
= \frac{1}{(\Sigma x_i^2)^2} E(x_1^2 \hat{u}_1^2 + \cdots + x_N^2 \hat{u}_N^2 + 2x_1 x_2 \hat{u}_1 \hat{u}_2 + \cdots + 2x_{N-1} x_N \hat{u}_{N-1} \hat{u}_N)
$$

$$
= \frac{1}{(\Sigma x_i^2)^2} [x_1^2 \sigma^2 + \cdots + x_N^2 \sigma^2] = \frac{\Sigma x_i^2}{(\Sigma x_i^2)^2} \sigma^2 = \frac{\sigma^2}{\Sigma x_i^2},
$$

where σ is the population residual variance. The sample residual variance $[S^2(\hat{u}) = \Sigma \hat{u}^2/(N-2)]$ is an unbiased estimate of the population residual variance. In Chapter 6, Problem 7, part (c), it is shown that if there is one explanatory variable,

$$
S^2(b_0) = \frac{S^2(\hat{u})}{N} + \bar{X}^2 S^2(b_1).
$$

Therefore

$$
S^2(b_0) = S^2(\hat{u})\left(\frac{1}{N} + \frac{\bar{X}^2}{\Sigma x^2}\right) = S^2(\hat{u})\left(\frac{\Sigma x^2 + N\bar{X}^2}{N \Sigma x^2}\right) = \frac{S^2(\hat{u})}{N}\left(\frac{\Sigma X^2}{\Sigma x^2}\right).
$$

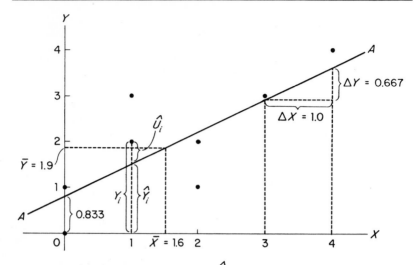

AA: Sample regression line, $\hat{Y}$ = 0.833 + 0.667X

FIG. 4. Scatter diagram of the data and the sample regression line.

and the square of the standard error of the intercept is

$$S^2(b_0) = \left(\frac{S^2(\hat{U})}{N}\right)\left(\frac{\sum\limits_{i=1}^{N} X_i^2}{\sum\limits_{i=1}^{N} x_i^2}\right),$$

where

$$\sum_{i=1}^{N} x_i^2 = \sum_{i=1}^{N} (X_i - \bar{X})^2.$$

To compute the residual variance $S^2(\hat{U})$, we have

$$y_i = b_1 x_i + \hat{U}_i,$$

$$\sum_{i=1}^{N} y_i^2 = b_1^2 \sum_{i=1}^{N} x_i^2 + \sum_{i=1}^{N} \hat{U}_i^2, \quad \text{since} \sum_{i=1}^{N} x_i \hat{u}_i = 0 \text{ (by assumption)},$$

$$\sum_{i=1}^{N} \hat{U}_i^2 = \sum_{i=1}^{N} y_i^2 - b_1^2 \sum_{i=1}^{N} x_i^2 = 12.90 - (0.667)^2(14.40) = 6.507,$$

$$S^2(\hat{U}) = \frac{\sum_{i=1}^{N} \hat{U}_i^2}{N - 2} = \frac{6.507}{8} = 0.813.$$

(Two degrees of freedom are lost in estimating the slope and intercept.) Then

$$S^2(b_1) = \frac{S^2(\hat{U})}{\sum_{i=1}^{N} x_i^2} = \frac{0.813}{14.4} = 0.0565,$$

$$S(b_1) = 0.238,$$

$$S^2(b_0) = \frac{S^2(\hat{U})}{N} \cdot \frac{\sum_{i=1}^{N} X_i^2}{\sum_{i=1}^{N} x_i^2} = \left(\frac{0.813}{10}\right)\left(\frac{40}{14.4}\right) = 0.225,$$

$$S(b_0) = 0.475.$$

Tests of Significance

Intercept

$$H_0:\beta_0 = 0, \qquad H_a:\beta_0 \neq 0.$$

$$t = \frac{b_0 - \beta_0}{S(b_0)} = \frac{0.833 - 0}{0.475} = 1.75.$$

For a 5 percent level of significance (type I error) and $df = N - 2 = 8$, the critical $t_{N-2} = 2.31$. The hypothesis that the intercept in the population is zero cannot be rejected. The intercept does not differ significantly from zero at a 5 percent level.

Slope

$$H_0 : \beta_1 = 0, \qquad H_a : \beta_1 \neq 0.$$

$$t = \frac{b_1 - \beta_1}{S(b_1)} = \frac{0.667 - 0}{0.238} = 2.80.$$

For a 5 percent level of significance and 8 degrees of freedom, the critical $t_{N-2} = 2.31$. The null hypothesis that the slope coefficient equals zero is rejected. The slope differs significantly from zero at a 5 percent level of significance.

(c) If the null hypothesis for the slope coefficient is $\beta_1 = 0$, then

$$t_{N-2} = \frac{b_1}{S(b_1)} = \frac{b_1 \sqrt{\sum_{i=1}^{N} x_i^2}}{S(\hat{U})} = \sqrt{\frac{b_1^2 \sum_{i=1}^{N} x_i^2}{S^2(\hat{U})}}$$

$$= \sqrt{\frac{(b^2 \sum_{i=1}^{N} x_i^2)/1}{(\sum_{i=1}^{N} \hat{U}_i^2)/(N-2)}} = \sqrt{\frac{(\sum_{i=1}^{N} \hat{y}_i^2)/1}{(\sum_{i=1}^{N} \hat{U}_i^2)/(N-2)}}$$

$$= \sqrt{F_{(1, N-2)}},$$

where F is the Fisher F-ratio (see Chapter 4). Recall that the F-distribution is the ratio of two independent chi-square distributions divided by their respective degrees of freedom. The predicted sum of squares ($\sum \hat{y}^2$) and the residual sum of squares ($\sum \hat{U}^2$) are statistically independent of each other and have chi-square distributions.

Note: Some computer programs print the F-ratio for each of the slope coefficients, but do not print the t-ratios. However, since $t_{N-2}^2 = F_{(1,N-2)}$, the t-ratio is easily computed. Whereas the t-test can be used to test the significance of only one explanatory variable, the F-test can be used to test the significance of a set of one or more explanatory variables (see Chapter 6, Problem 3).

(d) The elasticity of the dependent variable with respect to the independent variable is the percent change in the dependent variable associated with a 1 percent change in the explanatory variable. Unlike a slope, an elasticity is always a pure number, it has no units. A linear regression of Y on X assumes a constant slope, b_1. The elasticity of Y with respect to X at the point (X_i, Y_i) is

$$\varepsilon_{y,x} = \frac{\%\Delta Y_i}{\%\Delta X_i} = \frac{\Delta Y_i / Y_i}{\Delta X_i / X_i} = b_1 \frac{X_i}{Y_i}.$$

Computed at the mean the elasticity is

$$\varepsilon_{y,x} = b_1 \frac{\bar{X}}{\bar{Y}}.$$

For this problem the elasticity at the mean is

$$\varepsilon_{y,x} = b_1 \frac{\bar{X}}{\bar{Y}} = (0.667)\left(\frac{1.6}{1.9}\right) = 0.562.$$

(e) The regression equation can be reported as:

Variable	Regression coefficient	Standard error	Observed t-ratio
Constant	0.833	0.475	1.75
X	0.667	0.238	2.80

$S^2(\hat{U}) = 0.813$, $R^2 = 0.496$ (the term R^2 will be discussed below), $N = 10$, $df = 8$.

The regression results are often reported in the form of an equation with either the standard error or the t-ratio in parentheses below the regression coefficient:

$$\hat{Y} = 0.833 + 0.667X,$$
$$\quad\quad (1.75) \quad\quad (2.80)$$

$$S^2(\hat{U}) = 0.813, \quad R^2 = 0.496, \quad N = 10, \quad df = 8.$$

Problem 2. Hypothesis Testing of Regression Coefficients

Q: Using the data from Problem (1):

    (a) Is a unit increase in the desired number of children in the population accompanied by a unit increase in the actual number of children?

    (b) Is the actual number of children proportional to the desired number?

A: (a) The null hypothesis is that the slope coefficient is 1.0:

$$H_0: \beta_1 = 1, \quad H_a: \beta_1 \neq 1,$$

$$t = \frac{b_1 - \beta_1}{S(b_1)} = \frac{0.667 - 1.0}{0.238} = \frac{-0.333}{0.238} = -1.40.$$

    For a 5 percent level of significance, $df = 8$, the critical value is $t = 2.31$. The null hypothesis that the slope in the population (β_1) is unity cannot be rejected.

    (b) The dependent variable is proportional to the explanatory variable in the population if the population intercept equals zero. In Problem 1 the test of the statistical significance of the observed intercept indicated that we cannot reject the hypothesis $\beta_0 = 0$ at a 5 percent level of significance. $(t = (0.833 - 0.0)/0.475 = 1.75.)$ Thus we cannot reject the hypothesis of proportionality.

Problem 3. Coefficient of Determination

Q: Using the data from Problem (1):

    (a) To what extent are interfamily variations in the actual number of children explained by the desired number of children?

    (b) Adjust the explanatory power of the model for the small number of degrees of freedom.

A: (a) Regression analysis *cannot* be used to test causation. It is used to test for linear association or correlation. "Explanation" refers to correlation, not causation.

The model's (statistical) explanatory power is referred to as its *coefficient of determination*. Since

$$y_i = b_1 x_i + \hat{u}_i \qquad \text{and} \qquad \text{Cov }(x,\hat{u}) = 0,$$

it follows that

$$\underbrace{S^2(y)}_{\text{total variation}} = \underbrace{b_1^2 S^2(x)}_{\text{explained variation}} + \underbrace{S^2(\hat{u})}_{\text{residual variation}} .$$

The coefficient of determination (R^2) is the ratio of the variation in the dependent variable explained by the model, to the total variation in the dependent variable.

$$\begin{pmatrix} \text{Coefficient of} \\ \text{determination} \end{pmatrix} = R^2 = \frac{\text{explained variation}}{\text{total variation}} = \frac{b_1^2 S^2(x)}{S^2(y)} .$$

Then

$$R^2 = \frac{b_1^2 S^2(x)}{S^2(y)} = \left(\frac{\sum\limits_{i=1}^{N} x_i y_i}{\sum\limits_{i=1}^{N} x_i^2} \right)^2 \left(\frac{\sum\limits_{i=1}^{N} x_i^2}{\sum\limits_{i=1}^{N} y_i^2} \right)$$

$$= \left(\frac{\sum\limits_{i=1}^{N} x_i y_i}{\sqrt{\sum\limits_{i=1}^{N} x_i^2} \sqrt{\sum\limits_{i=1}^{N} y_i^2}} \right)^2 ,$$

and

$$R^2 = (\text{correlation between } x \text{ and } y)^2.$$

Since $y_i = \hat{y}_i + \hat{u}_i$ and it is assumed that

$$\sum_{i=1}^{N} \hat{y}_i \hat{u}_i = 0,$$

$$\sum_{i=1}^{N} y_i{}^2 = \sum_{i=1}^{N} \hat{y}_i{}^2 + \sum_{i=1}^{N} \hat{u}_i{}^2$$

and

$$R^2 = \frac{\displaystyle\sum_{i=1}^{N} \hat{y}_i{}^2}{\displaystyle\sum_{i=1}^{N} y_i{}^2} = 1 - \frac{\displaystyle\sum_{i=1}^{N} \hat{u}_i{}^2}{\displaystyle\sum_{i=1}^{N} y_i{}^2}.$$

Thus the coefficient of determination (R^2) is the square of the correlation coefficient between observed Y and predicted Y. For a simple regression this is the same as the square of the correlation between observed Y and observed X. In the example under examination

$$R^2 = 1 - \frac{6.5}{12.9} = 0.496,$$

since

$$\sum_{i=1}^{N} y_i{}^2 = 12.9 \qquad \text{and} \qquad \sum_{i=1}^{N} \hat{u}_i{}^2 = 6.5.$$

The explanatory (independent) variable explains approximately 50 percent of the variation in the dependent variable.

(b) If the sample size were made much smaller, R^2 would generally increase. For a sample size of two, $R^2 = 1$, yet this "perfect" explanatory power would have no meaning. It is desirable to have a measure of explanatory power that accounts for the number of degrees of freedom. The *adjusted coefficient of deter-*

mination (written as adj R^2 or $\bar{R}^2$) is†

$$\bar{R}^2 = 1 - \frac{\dfrac{\displaystyle\sum_{i=1}^{N} \hat{u}_i^2}{N-2}}{\dfrac{\displaystyle\sum_{i=1}^{N} y_i^2}{N-1}} = 1 - \left(\frac{N-1}{N-2}\right)\left(\frac{\displaystyle\sum_{i=1}^{N} \hat{u}_i^2}{\displaystyle\sum_{i=1}^{N} y_i^2}\right),$$

or

$$\bar{R}^2 = 1 - \left(\frac{N-1}{N-2}\right)(1 - R^2).$$

$\bar{R}^2$ is always smaller than R^2, but as the sample size (N) increases, $(N-1)/(N-2)$ approaches unity and $\bar{R}^2$ approaches R^2. Since $N = 10$, and $R^2 = 0.496$, we compute

$$\bar{R}^2 = 1 - \left(\frac{N-1}{N-2}\right)(1 - R^2) = 0.433.$$

The independent variable explains 43.3 percent of the variation in the dependent variable after adjustment is made for the small number of degrees of freedom.

The adjusted and unadjusted coefficients of determination can be used for descriptive purposes. However, tests of significance can be performed only for the square root of the unadjusted coefficient of determination, which is better known as the coefficient of correlation.

Problem 4. Coefficient of Correlation: Hypothesis Testing and Confidence Interval

Q: Using the data from Problem (1):

(a) Test the hypothesis that the correlation between X and Y is zero. Does this differ from the test of the statistical significance of the slope coefficient b_1?

† $\bar{R}^2$ is undefined for $N = 2$.

(b) Is the correlation coefficient in the population greater than 0.8?

(c) Construct a 95 percent confidence interval for the population correlation coefficient.

A: (a) The sampling distribution of the correlation coefficient (R) is known and is a function of the population value (ρ) and the number of degrees of freedom in the sample. However, the sampling distribution of the adjusted correlation coefficient $(\sqrt{\bar{R}^2})$ is not known. Hence tests of statistical significance for correlation coefficients are performed *only* on the unadjusted coefficient. The adjusted coefficient of determination $(\bar{R}^2)$ is used *only* for descriptive purposes.

If X and Y are normally distributed and the correlation coefficient of the population equals zero $(\rho_{xy} = 0)$, then $(R_{xy} - \rho_{xy})/S(R)$ has a t-distribution with $N - 2$ degrees of freedom because the sampling distribution of the numerator is normal. If $\rho = 0$, then

$$\rho^2 = \left(\frac{\beta_1^2 \sum_{i=1}^{N} x_i^2}{\sum_{i=1}^{N} y_i^2} \right) = 0 \qquad \text{and} \qquad \beta_1 = 0.$$

In a simple regression the test that the slope coefficient in the population equals zero is the same as the test that the correlation coefficient of the population equals zero. Since†

$$t^2 = \left(\frac{b_1}{S(b_1)} \right)^2 = \frac{b_1^2 \sum x_i^2}{\left(\sum \hat{U}_i^2 / (N - 2) \right)}$$

$$= (N - 2) \left(\frac{\sum y_i^2 - \sum \hat{U}_i^2}{\sum \hat{U}_i^2} \right)$$

$$= (N - 2) \left(\frac{1 - (\sum \hat{U}_i^2 / \sum y_i^2)}{(\sum \hat{U}_i^2 / \sum y_i^2)} \right)$$

$$= (N - 2) \left(\frac{R^2}{1 - R^2} \right) = \left(\frac{R_2}{\sqrt{\dfrac{1 - R^2}{N - 2}}} \right)^2 = \left(\frac{R}{S(R)} \right)^2$$

† For simplicity, $\sum\limits_{i=1}^{N}$ is replaced by $\sum$.

we have

$$S(R) = \sqrt{\frac{1 - R^2}{N - 2}}.$$

We previously found $R^2 = 0.496$, so that $R = 0.704$. Then

$$t = \frac{R}{\sqrt{\dfrac{1 - R^2}{N - 2}}} = \frac{0.704}{\sqrt{\dfrac{1.0 - 0.496}{8}}} = 2.80,$$

which is the same t-value as that obtained when testing the significance of the slope coefficient. The hypothesis that the correlation in the population (ρ) is zero is rejected.

Note: Tables exist for the sampling distribution of R when the population value equals zero for various degrees of freedom. (See Table 11 of the Appendix.) These tables permit a direct determination of the statistical significance of the sample correlation coefficient.
(b) If the population correlation coefficient (ρ) differs from zero, the sampling distribution of R is skewed and the t-test described above *cannot* be applied (see Figure 5). Fisher's Z-transformation changes the scale of R so that Z is approximately normally distributed even if ρ is near (plus or minus) unity. (*Note*: This

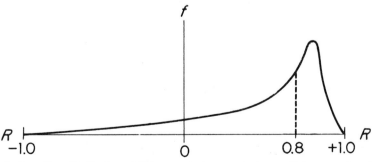

Fig. 5. Sampling distribution of R for population correlation coefficient equal to 0.8.

Z is *not* the same as the symbol Z used for the standardized normal variate.)

For the population,

$$Z = \frac{1}{2} \ln \left(\frac{1 + \rho}{1 - \rho} \right) ;$$

for the sample,

$$z = \frac{1}{2} \ln \left(\frac{1 + R}{1 - R} \right) ;$$

$$E(z) = Z \quad \text{and} \quad S(z) = \left(\frac{1}{\sqrt{N - k - 1}} \right) ,$$

The sample estimate z is an unbiased estimate of the population value (Z) and k is the number of estimated regression coefficients. k equals two in a simple regression. (Tables exist for the Z-transformation. See Table 12 in the Appendix.)

The null hypothesis is that the correlation coefficient in the population exceeds 0.8, and the alternative hypothesis is that this is not true:

$$H_0 : \rho > 0.8, \qquad H_a : \rho \leq 0.8.$$

The observed correlation is 0.704.

$$Z = \frac{1}{2} \ln \left(\frac{1 + 0.8}{1 - 0.8} \right) = 1.100,$$

$$z = \frac{1}{2} \ln \left(\frac{1 + 0.704}{1 - 0.704} \right) = 0.875.$$

Since the sampling distribution of z approaches normality,

$$t = \frac{z - Z}{S(z)} = \frac{0.875 - 1.100}{0.378} = -0.60,$$

where the Z-transform for 0.704 is 0.875 and for 0.8 it is 1.10 and $S(z) = 1/\sqrt{10 - 2 - 1} = 0.378$. For a 5 percent level of significance, $df = 8$, the one-tailed critical value is 1.86. The null hypothesis that the correlation coefficient of the population is greater than 0.8 cannot be rejected.

(c) The confidence interval for the correlation coefficient is

$$P\left(t_{\alpha/2} < \frac{z - Z}{S(z)} < t_{1-\alpha/2}\right) = 1 - \alpha.$$

For $\alpha = 0.05$,

$$P(z - t_{1-\alpha/2}S(z) < Z < z - t_{\alpha/2}S(z)) = 0.95.$$

For $R = 0.704$, we compute

$$z = \frac{1}{2}\ln\left(\frac{1 + R}{1 - R}\right) = 0.875,$$

$$S(z) = \frac{1}{\sqrt{N - k - 1}} = 0.378,$$

and for $\alpha/2 = 0.025$, the critical value is $t = 2.31$. Then,

$$P(0.01 < Z < 1.75) = 0.95.$$

Converting from Z to ρ, we get

$$P(0.01 < \rho < 0.94) = 0.95.$$

The confidence interval is very wide because the sample size is so small.

Problem 5. Reversing the Order of the Regression Equation

Q: Is the slope coefficient obtained from the regression of Y on X computed in Problem 1 the inverse of the slope coefficient obtained when X is regressed on Y?

A: If X is regressed on Y, $X_i = a_0 + a_1 Y_i + \hat{V}_i$, and

$$a_1 = \frac{\sum_{i=1}^{N} x_i y_i}{\sum_{i=1}^{N} y_i^2} = \frac{9.60}{12.90} = 0.74,$$

which does not equal the inverse of $b_1 = 0.667$ (that is, $0.74 \neq \frac{1}{0.667}$). For the regression of X on Y,

$$X_i = 0.19 + 0.74 Y_i + \hat{V}_i.$$

Then

$$a_1 b_1 = \left(\frac{\sum_{i=1}^{N} x_i y_i}{\sum_{i=1}^{N} y_i^2} \right) \left(\frac{\sum_{i=1}^{N} x_i y_i}{\sum_{i=1}^{N} x_i^2} \right) = \left(\frac{\sum_{i=1}^{N} x_i y_i}{\sqrt{\sum_{i=1}^{N} y_i^2} \sqrt{\sum_{i=1}^{N} x_i^2}} \right)^2 = R^2,$$

where R is the correlation between X and Y. $b_1 = 1/a_1$ only in the case in which all the observations fall on a straight line ($R^2 = 1$).

Both regression lines go through the mean of the data (see Figure 6). In the regression of Y on X ($Y_i = b_0 + b_1 X_i + \hat{U}_i$), the

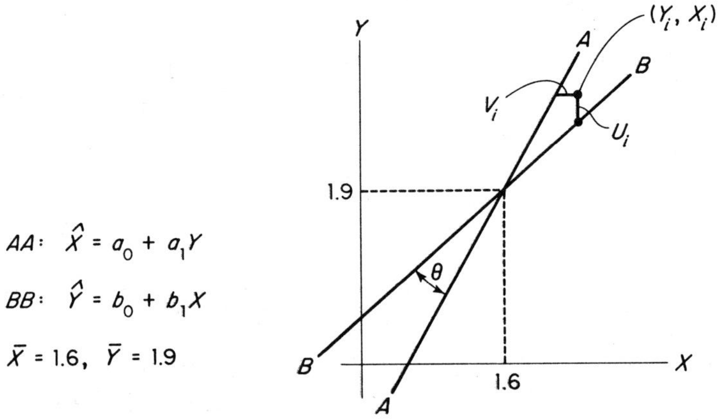

AA: $\hat{X} = a_0 + a_1 Y$

BB: $\hat{Y} = b_0 + b_1 X$

$\bar{X} = 1.6$, $\bar{Y} = 1.9$

FIG. 6

sum of the squared vertical deviations of the observations from the regression line is minimized. In the regression of X on Y ($X_i = a_0 + a_1 Y_i + \hat{V}_i$), the sum of the squared horizontal deviations of the observations from the regression line is minimized. If all the observations fall on a straight line, there are no deviations to be minimized and the same line is obtained whether X is regressed on Y or Y is regressed on X. If the correlation between X and Y is zero, $b_1 = 0$ and $a_1 = 0$ and the regression equation of Y on X is $\hat{Y} = b_0$, while the regression of X on Y is $\hat{X} = a_0$ (see Figure 7). Thus the angle (θ) between the two regression lines at the mean is at a minimum value of zero degrees ($0°$) when $R^2 = 1$ and at a maximum of $90°$ when $R^2 = 0$.

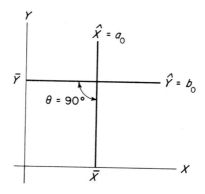

FIG. 7. Regression lines if X and Y are not correlated.

Problem 6. Heteroscedastic Residuals

Q: Using the data in Problem (1): (a) Test the hypothesis that the residual variance depends on the level of the explanatory variable X.

(b) What problems arise if the residual variance is not constant for all values of X?

(c) What techniques may be used to correct for heteroscedastic residuals?

(d) When does heteroscedasticity arise if sample means are used for the value of the dependent variable?

A: (a) The residuals are homoscedastic if the residual variance is the same for all values of the explanatory variable, and the residuals are heteroscedastic if the residual variance depends on the value of the explanatory variable. To test for heteroscedasticity, divide the explanatory variable into intervals and compute the residual variance for each interval. Then test for the equality of variances (see Chapter 4). If more than two intervals are to be used, it is easiest to construct intervals with (approximately) the same number of observations so that the F_{max} test can be applied (see Chapter 4).

The predicted values of the dependent variable ($\hat{Y}$) are obtained from the regression equation on p. 146.

	Intervals for X	X	Y	$\hat{Y}$	$Y - \hat{Y} = \hat{U}$	$\hat{U}^2$
(A)	0–1	0	0	0.833	−0.833	0.694
		1	2	1.500	0.500	0.250
		1	3	1.500	1.500	2.250
		0	1	0.833	0.167	0.028
		1	2	1.500	0.500	0.250
						3.472
(B)	2–4	2	1	2.167	−1.167	1.362
		3	3	2.834	0.166	0.028
		4	4	3.501	0.499	0.249
		2	2	2.167	−0.167	0.028
		2	1	2.167	−1.167	1.362
						3.029

The residual variance in each interval is

$$S_A^2(\hat{U}) = \frac{3.472}{5}, \qquad S_B^2(\hat{U}) = \frac{3.029}{5}.$$

To test for equality of variances, we have

$$H_0 : \sigma_A^2 = \sigma_B^2, \qquad H_a : \sigma_A^2 \neq \sigma_B^2.$$

observed $F_{(\eta_A, \eta_B)} = \dfrac{S_A^2 \, \sigma_B^2}{S_B^2 \, \sigma_A^2} = \dfrac{3.472}{3.029} = 1.15$

if the null hypothesis $\sigma_A^2 = \sigma_B^2$ is correct, and η_A and η_B are the degrees of freedom. The critical F-value for a 5 percent type I (α) error is $F_{(4,4)} = 6.39$. The null hypothesis of equal variances cannot be rejected. The residuals are consistent with the hypothesis of equal variances.

(b) Heteroscedasticity does not bias the estimate of regression coefficients. It may, however, bias the estimated standard error of the regression slope coefficient. For example, if the residual variance increases with the value of X, the standard error of the regression slope coefficient is biased upward. If the standard error is biased upward, but the coefficient is not statistically significant, it is not certain whether it would be significant for an unbiased standard error.

   In economic problems a positive relationship between the residual variance and the explanatory variable is far more common than a negative relationship. For instance, studies of the effect of income (I) on consumption expenditures (C) indicate that the variance of consumption expenditures within income classes increases for higher levels of income.

   If there is heteroscedasticity the standard error of prediction of the dependent variable Y for a given value of X depends on the residual variance for *that* value of X. Acceptance of the assumption of homoscedasticity permits an easy computation of standard errors for the predicted values of the dependent variable (see Problem 7).

(c) There is a simple transformation which is often used successfully in correcting for heteroscedasticity that arises when the variance of the dependent variable increases with the value of the explanatory variable. Suppose $Y_i = b_0 + b_1 X_i + \hat{U}_i$ and $S(\hat{U}_i) = S(Y_i \mid X_i) = kX_i$, where $k \neq 0$. The residual variance from this regression would not be constant for all values of X. However, if we divide both sides of the regression equation by X_i we obtain

$$\frac{Y_i}{X_i} = \frac{b_0}{X_i} + b_1 + \frac{\hat{U}_i}{X_i}.$$

Computing the regression gives

$$\frac{Y_i}{X_i} = a_0 + a_1 \frac{1}{X_i} + \hat{V}_i.$$

Then $\hat{V}_i = \hat{U}_i/X_i$, a_0 is the estimate of b_1, and a_1 is the estimate of b_0. The residual variance is

$$S^2(\hat{V}_i) = \frac{S^2(\hat{U}_i)}{X_i^2} = \frac{k^2 X_i^2}{X_i^2} = k^2,$$

which is a constant.

Using this procedure we transformed a regression equation with a heteroscedastic residual to a form with a homoscedastic residual. However, if the standard deviation from the regression increases but does not increase in direct proportion to the explanatory variable, this procedure may reduce, but need not eliminate, heteroscedasticity. For example, suppose $S^2(\hat{U}_i) = k_0 + k_1 X_i^2$. Then if $\hat{V}_i = \hat{U}_i/X_i$,

$$S^2(\hat{V}_i) = \frac{k_0 + k_1 X_i^2}{X_i^2} = \frac{k_0}{X_i^2} + k_1.$$

The extent of heteroscedasticity in the original and transformed regression depends on the relative sizes of k_1 and k_0.

(d) Heteroscedastic residuals may also arise when the dependent variable is a sample mean. Suppose that if income for an individual Y is regressed on X the residual $S^2(\hat{U})$ is homoscedastic. Suppose the dependent variable under investigation is mean income in a region $(\bar{Y}_i)$ and the sample size is N_i in the ith region. Then since $\bar{Y}_i = b_0 + b_1 X_i + \bar{U}_i$,

$$S^2(\bar{U}_i) = \frac{S^2(\hat{U}_i)}{N_i}$$

and the residuals are heteroscedastic if the sample sizes differ substantially across regions.

This form of heteroscedasticity can be corrected by computing a *weighted regression*, where the weight for the ith observation is the square root of the sample size $(\sqrt{N_i})$. If the regression is

$$\boxed{\sqrt{N_i}(\bar{Y}_i) = b_0 + b_1(\sqrt{N_i}\,X_i) + \hat{V}_i,}$$

then for the observations in the ith region,

$$\hat{V}_i = \sqrt{N_i}\,\bar{U}_i$$

and

$$S^2(\hat{V}_i) = N_i S^2(\bar{U}_i) = \frac{N_i S^2(\hat{U}_i)}{N_i} = S^2(\hat{U}_i).$$

The residuals are now homoscedastic.

Weighted regressions that are used to correct for heteroscedastic residuals are often called *generalized least-squares regressions*.

Problem 7. Prediction from a Simple Regression

Q: (a) An eleventh newly married couple indicated that they desired to have three children. On the basis of the regression analysis for the first ten couples (Problem 1), predict the number of children actually had by the eleventh couple.

(b) The eleventh couple actually had no children. Is this consistent with the hypothesis that this couple comes from the same "population" as the couples that comprised the sample?

(c) Construct a 95 percent confidence interval for the mean number of children had by couples who reported that they desired to have three children.

A: The procedure for computing prediction intervals assumes homoscedastic residuals. The preceding problem indicates that the hypothesis of equal variances for different values of the explanatory variable cannot be rejected for the data under investigation.

It would be inappropriate to predict values of the dependent variable for values of the explanatory variable that are outside the range of the data for the explanatory variable. The regression equation is the "best" linear fit to the data. If the true relation is nonlinear, predictions outside the range of the data may result in large but unknown prediction errors (see Figure 8). In this problem, the value of the explanatory variable used for prediction is within the range of the data.

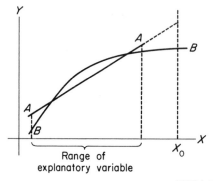

AA: Regression line computed from data

BB: True nonlinear relation

X_0: Value of explanatory variable outside of range of data

Range of explanatory variable

X_0

FIG. 8

(a) The regression equation computed in Problem 1 is

$$\hat{Y}_i = 0.833 + 0.667X_i.$$

The mean predicted number of children actually had for $X_{N+1} = 3$ is $Y_p = 2.834$, or 3 to the nearest integer.

(b) To test the hypothesis that the eleventh family's behavior is consistent with the regression equation computed in Problem 1 we construct an observed t-ratio:

$$t = \frac{Y_p - Y_{N+1}}{S_p},$$

where Y_{N+1} is the observed value of the dependent variable for the $N + 1$-st observation, Y_p is the predicted value and S_p is the standard error of prediction. The observed value is $Y_{N+1} = 0$; the predicted value is $Y_p = 2.834$.

To obtain the standard error of the predicted value of the dependent variable for the $N + 1$-st observation, we use

$$Y_p = b_0 + b_1 X_{N+1} + \hat{U}_{N+1} = \bar{Y} + b_1 x_{N+1} + \hat{U}_{N+1}.$$

Then, computing the variance of both sides of the equation,

$$S_p{}^2 = S^2(\bar{Y}) + x_{N+1}{}^2 S^2(b_1) + S^2(\hat{U}).$$

The covariance terms are all zero. The sampling variability of the predicted value of Y has been divided into the statistically independent effects of sampling variation on

(i) the level of the regression line $[S^2(\bar{Y})]$,
(ii) the slope of the regression line $[S^2(b_1)]$,

(iii) the residual variation $[S^2(\hat{U})]$.

For a given value of X,

$$S^2(\bar{Y}) = \frac{S^2(\hat{U})}{N}.$$

Also

$$S^2(b_1) = \frac{S^2(\hat{U})}{\sum\limits_{i=1}^{N} x_i^2}.$$

Then

$$S_p^2 = \frac{S^2(\hat{U})}{N} + x_{N+1}^2 \left(\frac{S^2(\hat{U})}{\sum\limits_{i=1}^{N} x_i^2} \right) + S^2(\hat{U})$$

$$S_p^2 = S^2(\hat{U}) \left(1 \quad + \quad \frac{1}{N} \quad + \quad \frac{x_{N+1}^2}{\sum\limits_{i=1}^{N} x_i^2} \right).$$

| Effect of | residual variation | level error | slope error |

Since $x_{N+1}^2 = (X_{N+1} - \bar{X})^2$, S_p^2 is larger the further the value of the explanatory variable is from the mean. The standard error of prediction (S_p) is smallest at the mean $(X_{N+1} = \bar{X})$ because the regression line goes through the mean of the data and at the mean there is no slope error (see Figure 9).

Using the data from p. 145 and p. 149,

$$S_p^2 = (0.813)\left(1 + \frac{1}{10} + \frac{(3.0 - 1.6)^2}{14.40}\right) = 1.005,$$

$$S_p = \sqrt{1.005} = 1.002.$$

$$\text{observed } t = \left(\frac{Y_p - Y_{N+1}}{S_p}\right) = \frac{2.8 - 0.0}{1.002} = 2.8.$$

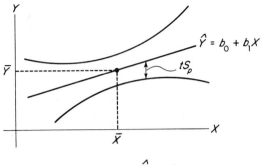

Acceptance interval = $\hat{Y} \pm tS_p$

FIG. 9

For $df = 8$ and $\alpha = 0.05$, the critical $t = 2.31$. The null hypothesis that the eleventh couple comes from the same population as the sample of ten couples is rejected under a 5 percent level of significance.

(c) The mean number of children predicted for couples reporting they expect to have three children is $Y_p = 2.834$. Since the mean residual equals zero for a given value of X, the standard error of prediction for the mean value Y_p is computed from

$$Y_p = b_0 + b_1 X_{N+1} = \bar{Y} + b_1 x_{N+1}.$$

$$S_{\bar{p}}^2 = S^2(\bar{Y}) + x_{N+1}^2 S^2(b_1) = S^2(\hat{U})\left(\frac{1}{N} + \frac{x_{N+1}^2}{\displaystyle\sum_{i=1}^{N} x_i^2} \right)$$

| Effect of | Level error | Slope error | Level error | Slope error |

(There is no residual error for the *mean* predicted value.)

$$S_{\bar{p}}^2 = (0.813)\left(\frac{1}{10} + \frac{(3.0 - 1.6)^2}{14.40} \right) = 0.192,$$

$$S_{\bar{p}} = 0.44.$$

The confidence interval is

$$P\left(-t < \frac{Y_p - \bar{Y}_{N+1}}{S_{\tilde{p}}} < +t\right) = 1 - \alpha,$$

$$P(Y_p - tS_{\tilde{p}} < \bar{Y}_{N+1} < Y_p + tS_{\tilde{p}}) = 1 - \alpha.$$

The critical value of t is

$$t_{N-2}(\alpha = 0.05) = 2.31,$$

and thus

$$P\big(2.834 - (2.31)(0.44) < \bar{Y}_{N+1}$$
$$< 2.834 + (2.31)(0.44)\big) = 0.95,$$
$$P(1.82 < \bar{Y}_{N+1} < 3.85) = 0.95.$$

There is a 95 percent probability that the true mean number of children born to those desiring three children is in the interval 1.82 to 3.85 children.

Note that the standard error of prediction is smaller for the mean predicted value of the dependent variable ($S_{\tilde{p}} = 0.44$) than for the predicted value of the dependent variable ($S_p = 1.002$).

Problem 8. Functional Form

Q: (a) Is the linear functional relationship between Y and X in Problem 1 the only functional form that could have been used? What criteria influence the selection of the functional form?

(b) Compute the equation for the regression of Y on X^2. Compare the coefficient of determination, the slope of Y on X, and the elasticity of Y on X for this regression with the linear regression of Y on X.

A: (a) In performing a regression analysis the investigator must select a function to relate the dependent variable to the explanatory variable. Three principles should influence the selection of the function. First, the economic model that is motivating the empirical analysis may suggest a particular structure for the equation. Second, a simple functional form is preferred to one that is more complicated. Third, a functional form that is a better fit to the data is preferred to one that offers a poorer fit. The two most common functional forms are the linear and double-log equations, and these are preferred for their simplicity.

Some functional forms are:†

Name	Equation	Remarks
Linear	$Y = a + bX + \hat{U}$	Constant slope $= b$
Semi-log	$\ln Y = a + bX + \hat{U}$	
Semi-log	$Y = a + b \ln X + \hat{U}$	
Double-log	$\ln Y = a + b \ln X + \hat{U}$	Constant elasticity $= b$
Hyperbolic	$Y = a + b(1/X) + \hat{U}$	As X approaches infinity Y approaches the constant a.
Parabolic (quadratic)	$Y = a + bX + cX^2 + \hat{U}$	Computed as if it were a multiple regression with two explanatory variables.

For the *same* dependent variable, a functional form provides a "better" fit if it has a higher adjusted R^2. A comparison of $\bar{R}^2$ should not be made to determine which functional form provides the best fit if the dependent variables differ (e.g., Y, Y^2 and $\ln Y$ are three *different* dependent variables), since the models would be explaining variation for different variables.

(b) We wish to compute the regression

$$Y_i = C_0 + C_1(X_i^2) + \hat{V}_i.$$

Let us define $Z = X^2$. Then

$$Y_i = C_0 + C_1 Z_i + \hat{V}_i$$

† An additional functional form, the logistic function, is discussed in Problem 8, Chapter 6.

is a linear regression equation, where (see p. 146)

$$C_1 = \frac{\sum\limits_{i=1}^{N} y_i z_i}{\sum\limits_{i=1}^{N} z_i^2} \quad \text{and} \quad C_0 = \bar{Y} - C_1 \bar{Z}.$$

X_i	$X_i^2 = Z_i$	$Z_i - \bar{Z} = z_i$	z_i^2	Y_i	$Y_i - \bar{Y} = y_i$	$z_i y_i$	y_i^2
0	0	-4	16	0	-1.9	7.6	3.61
1	1	-3	9	2	0.1	-0.3	0.01
2	4	0	0	1	-0.9	0.0	0.81
1	1	-3	9	3	1.1	-3.3	1.21
0	0	-4	16	1	-0.9	$+3.6$	0.81
3	9	5	25	3	1.1	5.5	1.21
4	16	12	144	4	2.1	25.2	4.41
2	4	0	0	2	0.1	0.0	0.01
2	4	0	0	1	-0.9	0.0	0.81
1	1	-3	9	2	0.1	-0.3	0.01
Sum: 16	40	0	228	19	0.0	38.0	12.90

$$N = 10, \quad \bar{Z} = \frac{\sum\limits_{i=1}^{N} Z_i}{N} = 4.0, \quad \bar{Y} = \frac{\sum\limits_{i=1}^{N} Y_i}{N} = 1.9,$$

$$C_1 = \frac{\sum\limits_{i=1}^{N} y_i z_i}{\sum\limits_{i=1}^{N} z_i^2} = \frac{38.0}{228} = 0.167,$$

$$C_0 = \bar{Y} - C_1 \bar{Z} = 1.9 - (0.167)(4.0) = 1.233.$$

The linear regression equation of Y on X^2 is

$$\hat{Y}_i = 1.233 + 0.167 X_i^2.$$

We now wish to compute the explanatory power of the model

and the standard errors of the regression coefficients. The residual variance can be computed as follows:

$$\sum_{i=1}^{N} y_i^2 = C_1^2 \sum_{i=1}^{N} z_i^2 + \sum_{i=1}^{N} \hat{V}_i^2,$$

$$\sum_{i=1}^{N} \hat{V}_i^2 = \sum_{i=1}^{N} y_i^2 - C_1^2 \sum_{i=1}^{N} z_i^2 = 12.9 - (0.167)^2(228),$$

$$\sum_{i=1}^{N} \hat{V}^2 = 6.56.$$

$$S^2(\hat{V}) = (\sum_{i=1}^{N} \hat{V}_i^2)/(N - 2) = \frac{6.56}{8} = 0.82$$

is the residual variance. The explanatory power of the model is

$$R^2 = 1 - \frac{\displaystyle\sum_{i=1}^{N} \hat{V}_i^2}{\displaystyle\sum_{i=1}^{N} y_i^2} = 1 - \frac{6.56}{12.9} = 0.492.$$

Computing the standard error of the slope coefficient, we have

$$S^2(C_1) = \frac{S^2(\hat{V})}{\displaystyle\sum_{i=1}^{N} z_i^2} = \frac{\displaystyle\sum_{i=1}^{N} \hat{V}_i^2}{(N - 2) \displaystyle\sum_{i=1}^{N} z_i^2} = \frac{6.56}{(8)(228)} = 0.0035,$$

$$S(C_1) = 0.060.$$

Computing the standard error of the intercept,

$$S^2(C_0) = \frac{S^2(\hat{V})}{N} \frac{\displaystyle\sum_{i=1}^{N} Z_i^2}{\displaystyle\sum_{i=1}^{N} z_i^2} = \frac{(0.82)(388)}{(10)(228)} = 0.140,$$

$$S(C_0) = 0.374.$$

The regression of Y on X^2 can be reported as:

Variable	Regression coefficient	Standard error	Observed t-ratio
Constant	1.233	0.374	3.30
X^2	0.167	0.060	2.78

where $S^2(\hat{V}) = 0.82$, $R^2 = 0.492$, $N = 10$, and $df = 8$ (see Figure 10).

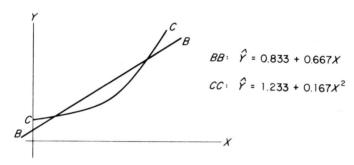

BB: $\hat{Y} = 0.833 + 0.667X$

CC: $\hat{Y} = 1.233 + 0.167X^2$

FIG. 10. Graphic representation of the regression of Y on X and Y on X^2.

The slope of the regression line $\hat{Y} = C_0 + C_1X^2$ is

$$\frac{\Delta Y}{\Delta X} = 2(C_1X),$$

which at the mean X ($\bar{X} = 1.6$) is

$$\frac{\Delta Y}{\Delta X} = 2(0.167)(1.6) = 0.534.$$

The elasticity at the mean X is

$$\varepsilon_{y,x} = \frac{\%\Delta Y}{\%\Delta X} = \frac{\Delta Y}{\Delta X}\frac{\bar{X}}{\bar{Y}} = (0.534)\left(\frac{1.6}{1.9}\right) = 0.45.$$

Results from regression of Y on X and X^2 are:

	X†	X^2
R^2	0.496	0.492
Slope‡	0.667	0.534
Elasticity‡	0.562	0.450

† Obtained from Problem 1.
‡ Y on X, at the mean $(\bar{Y}, \bar{X})$.

Problem 9. Errors in Variables

Q: A theory predicts a linear relationship between the true value of the variable Y and the true value of the variable X: $Y_t = \beta_0 + \beta_1 X_t + U_t$.

It is known, however, that errors occur in the measurement of variables X and Y. Could these errors bias the estimate of the regression coefficients?

A: The true relationship in the population is $Y_t = \beta_0 + \beta_1 X_t + U_t$. The observed (measured) values of the variables are $Y = Y_t + v$ and $X = X_t + e$, where v and e are measurement errors that are assumed to have zero means, that is, $\bar{v} = \bar{e} = 0$. The observed regression equation is

$$Y_i = b_0 + b_1 X_i + \hat{U}_i.$$

The computed slope coefficient† is

$$
b_1 = \frac{\sum xy}{\sum x^2} = \frac{\sum (x_t + e)(y_t + v)}{\sum (x_t + e)^2}
$$

$$
= \frac{\sum x_t y_t + \sum x_t v + \sum y_t e + \sum ev}{\sum x_t^2 + \sum e^2 + 2 \sum x_t e}.
$$

† $\displaystyle\sum_{i=1}^{N}$ is replaced by $\sum$ and the i subscript is suppressed.

If the measurement error is purely random, the error terms are uncorrelated (zero covariance) with the true value of the explanatory variables and with each other:

$$\sum x_t e = \sum y_t e = \sum e v = \sum x_t v = 0.$$

Then

$$b_1 = \frac{\sum x_t y_t}{\sum x_t^2 + \sum e^2} \quad \text{and } b_1 \text{ is less than} \quad \frac{\sum x_t y_t}{\sum x_t^2}.$$

If there is purely random error in measuring the explanatory variable, the observed regression slope coefficient is a downward biased estimate of the true regression coefficient. If the slope coefficient is biased downward, the intercept is biased upward since $b_0 = \bar{Y} - b_1 \bar{X}$ (see Figure 11).

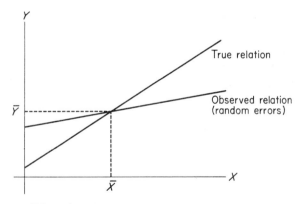

Fig. 11. Effect of random measurement error on the regression line.

If the variables in a regression are subject to random measurement error, a variable that has a significant observed slope coefficient will necessarily have a significant coefficient using the true values. Purely random measurement error can result in an insignificant observed slope coefficient even if the relationship would be significant if there were no error.

If random measurement errors exist in the dependent variable but there are no measurement errors in the explanatory variable, the slope coefficient is not biased. If $e_i = 0$ for all i and $\sum x_i v = 0$, then

$$b_1 = \frac{\sum xy}{\sum x^2} = \frac{\sum x_i y_i}{\sum x_t^2}$$

and there is no bias.

If measurement errors are not purely random, the slope coefficient can be biased upward. For example, suppose we are interested in regressing income on schooling and there is no error in reporting level of schooling ($X_t = X$, and $e = 0$). However, persons with higher levels of schooling have higher incomes (Y_t) and tend to over-report their income, while persons with less schooling tend to under-report their income. That is, there is a positive covariance between X_t and Y_t and between X_t and v (the error in measuring Y_t). Then, if

$$Y_i = b_0 + b_1 X_i + \hat{U}_i,$$

it follows that

$$b_1 = \frac{\sum x_i y_i + \sum x_i v}{\sum x_t^2} > \frac{\sum x_i y_i}{\sum x_t^2}.$$

The positive correlation of the measurement error in income with schooling ($\sum x_i v > 0$) biases upward the slope coefficient of schooling.

When there are substantial random errors of measurement in the explanatory variable in a very large sample, *instrumental variables* may be used to compute unbiased estimates of the regression coefficients. The instrumental variables technique is discussed in Problem 5, Chapter 7.

Chapter 6

Multiple Regression Analysis

The general principles of regression analysis were developed in Chapter 5 for the case of one explanatory variable. A *multiple regression* has more than one explanatory variable. For the case of two explanatory variables the population equation is written as

$$Y_i = \beta_0 + \beta_1 X_{1,i} + \beta_2 X_{2,i} + U_i,$$

where $X_{1,i}$ and $X_{2,i}$ are the values of the two explanatory variables for the ith observation. Figure 1 is a graphic presentation of the population regression plane in a three-dimensional space for a two explanatory variable model. The plane is drawn under the assumption that the intercept is positive ($\beta_0 > 0$) and the two slope coefficients are negative ($\beta_1 < 0, \beta_2 < 0$). The multiple regression plane goes through the mean point of the variables, $P(\bar{Y}, \bar{X}_1, \bar{X}_2)$.

The sample regression equation is written as

$$Y_i = b_0 + b_1 X_{1,i} + b_2 X_{2,i} + \hat{U}_i.$$

The decision rule for obtaining the sample estimates of the regression coefficients is to find the values that maximize the explanatory power

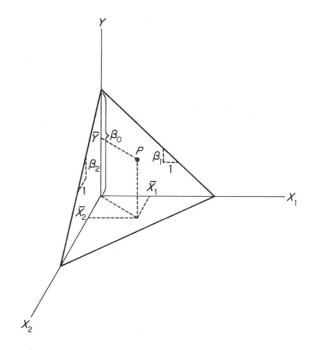

1. Equation of regression plane:

$$Y_i = \beta_0 + \beta_1 X_{1,i} + \beta_2 X_{2,i} + U_i$$

2. Drawn under assumption:

$$\beta_0 > 0, \quad \beta_1 < 0, \quad \beta_2 < 0$$

3. P: Point on plane at mean $\bar{Y}, \bar{X}_1, \bar{X}_2$

FIG. 1. Population regression plane—two explanatory variable model.

(coefficient of determination) of the regression model. This is the same as finding the regression coefficients that minimize the sum of the squared residual, $\hat{U}_i$. In the two explanatory variable case the residual is

$$\hat{U}_i = Y_i - b_0 - b_1 X_{1,i} - b_2 X_{2,i}.$$

Squaring both sides of the equation and summing across all N observations gives

$$\sum_{i=1}^{N} \hat{U}_i^2 = \sum_{i=1}^{N} (Y_i - b_0 - b_1 X_{1,i} - b_2 X_{2,i})^2.$$

Setting the first derivatives of the sum of squared residuals with respect to the regression coefficients equal to zero, we have

$$\frac{\partial \sum_{i=1}^{N} \hat{U}_i^2}{\partial b_0} = 2 \sum_{i=1}^{N} (Y_i - b_0 - b_1 X_{1,i} - b_2 X_{2,i})(-1) = 0,$$

$$\frac{\partial \sum_{i=1}^{N} \hat{U}_i^2}{\partial b_1} = 2 \sum_{i=1}^{N} (Y_i - b_0 - b_1 X_{1,i} - b_2 X_{2,i})(-X_{1,i}) = 0,$$

$$\frac{\partial \sum_{i=1}^{N} \hat{U}_i^2}{\partial b_2} = 2 \sum_{i=1}^{N} (Y_i - b_0 - b_1 X_{1,i} - b_2 X_{2,i})(-X_{2,i}) = 0.$$

Then the normal equations can be written as

$$\bar{Y} = b_0 + b_1 \bar{X}_1 + b_2 \bar{X}_2,$$

$$\sum_{i=1}^{N} Y_i X_{1,i} = b_0 \sum_{i=1}^{N} X_{1,i} + b_1 \sum_{i=1}^{N} X_{1,i}^2 + b_2 \sum_{i=1}^{N} X_{1,i} X_{2,i},$$

$$\sum_{i=1}^{N} Y_i X_{2,i} = b_0 \sum_{i=1}^{N} X_{2,i} + b_1 \sum_{i=1}^{N} X_{1,i} X_{2,i} + b_2 \sum_{i=1}^{N} X_{2,i}^2.$$

Using lower-case letters to designate deviations from means ($y_i = Y_i - \bar{Y}$, $x_{1,i} = X_{1,i} - \bar{X}_1$, $x_{2,i} = X_{2,i} - \bar{X}_2$), we can write the three normal equations as

$$\bar{Y} = b_0 + b_1\bar{X}_1 + b_2\bar{X}_2,$$

$$\sum_{i=1}^{N} y_i x_{1,i} = b_1 \sum_{i=1}^{N} x_{1,i}^2 + b_2 \sum_{i=1}^{N} x_{1,i} x_{2,i},$$

$$\sum_{i=1}^{N} y_i x_{2,i} = b_1 \sum_{i=1}^{N} x_{1,i} x_{2,i} + b_2 \sum_{i=1}^{N} x_{2,i}^2.$$

There is always one normal equation for each coefficient in the estimated multiple regression. These equations can be solved for the values of the regression coefficients if the explanatory variables, or a subset of the explanatory variables, are not perfectly or nearly perfectly correlated with each other. Problems arise when one explanatory variable is very highly correlated with one or more other variables. This situation is called *multicollinearity*.†

If the two explanatory variables, X_1 and X_2, are not perfectly correlated $[R^2(X_1,X_2) \neq 1]$, solving the normal equations for b_0, b_1, and b_2,‡

† The inability to obtain unique solutions when there is perfect multicollinearity is most easily shown in the two explanatory variable model. Suppose X_1 is perfectly correlated with X_2, that is, $R^2(X_1, X_2) = 1$. Then X_2 can be written as a linear function of X_1: $X_2 = c_0 + c_1 X_1$. Substituting, the three normal equations become

$$\sum_{i=1}^{N} y_i x_{1,i} = b_1 \sum_{i=1}^{N} x_{1,i}^2 + (b_2 c_1) \sum_{i=1}^{N} x_{1,i}^2,$$

$$\sum_{i=1}^{N} y_i x_{1,i} = b_1 \sum_{i=1}^{N} x_{1,i}^2 + (b_2 c_1) \sum_{i=1}^{N} x_{1,i}^2,$$

$$\bar{Y} = (b_0 + c_0) + b_1\bar{X}_1 + (b_2 c_1)\bar{X}_1.$$

The first and second equations are identical. There are now only two equations and three unknowns, b_0, b_1, and b_2. Without prior information about the value of one of the regression coefficients or a relationship among the coefficients it is not possible to obtain unique solutions for the regression coefficients.

$$\ddagger \sum_{i=1}^{N} x_{1,i}^2 \sum_{i=1}^{N} x_{2,i}^2 - \left(\sum_{i=1}^{N} x_{1,i} x_{2,i}\right) = \left[\sum_{i=1}^{N} x_{1,i}^2 \sum_{i=1}^{N} x_{2,i}^2\right] \left[1 - \frac{\left(\sum_{i=1}^{N} x_{1,i} x_{2,i}\right)^2}{\sum_{i=1}^{N} x_{1,i}^2 \sum_{i=1}^{N} x_{2,i}^2}\right]$$

equals zero (0)

for $R^2(X_1,X_2) = 1$ and equals $\sum_{i=1}^{N} x_{1,i}^2 \sum_{i=1}^{N} x_{2,i}^2$ for $R^2(X_1,X_2) = 0$.

we have

$$b_0 = \bar{Y} - b_1\bar{X}_1 - b_2\bar{X}_2,$$

$$b_1 = \dfrac{\displaystyle\sum_{i=1}^{N} x_{1,i}y_i \sum_{i=1}^{N} x_{2,i}{}^2 - \sum_{i=1}^{N} x_{1,i}x_{2,i} \sum_{i=1}^{N} x_{2,i}y_i}{\displaystyle\sum_{i=1}^{N} x_{1,i}{}^2 \sum_{i=1}^{N} x_{2,i}{}^2 - (\sum_{i=1}^{N} x_{1,i}x_{2,i})^2},$$

$$b_2 = \dfrac{\displaystyle\sum_{i=1}^{N} x_{2,i}y_i \sum_{i=1}^{N} x_{1,i}{}^2 - \sum_{i=1}^{N} x_{1,i}x_{2,i} \sum_{i=1}^{N} x_{1,i}y_i}{\displaystyle\sum_{i=1}^{N} x_{1,i}{}^2 \sum_{i=1}^{N} x_{2,i}{}^2 - (\sum_{i=1}^{N} x_{1,i}x_{2,i})^2}.$$

The standard errors and the covariance of these regression coefficients can be computed from†

$$S^2(b_1) = S^2(\hat{U})\left(\dfrac{\displaystyle\sum_{i=1}^{N} x_{2,i}{}^2}{\displaystyle\sum_{i=1}^{N} x_{1,i}{}^2 \sum_{i=1}^{N} x_{2,i}{}^2 - (\sum_{i=1}^{N} x_{1,i}x_{2,i})^2}\right),$$

$$S^2(b_2) = S^2(\hat{U})\left(\dfrac{\displaystyle\sum_{i=1}^{N} x_{1,i}{}^2}{\displaystyle\sum_{i=1}^{N} x_{1,i}{}^2 \sum_{i=1}^{N} x_{2,i}{}^2 - (\sum_{i=1}^{N} x_{1,i}x_{2,i})^2}\right),$$

† Using the previous footnote,

$$S^2(b_1) = S^2(\hat{U})\left(\dfrac{1}{\displaystyle\sum_{i=1}^{N} x_{1,i}{}^2(1 - R^2)}\right).$$

Ceteris paribus, $S^2(b_1)$ increases as $R^2(X_1,X_2)$ increases, and decreases as N increases.

$$\text{Cov } (b_1, b_2) = S^2(\hat{U}) \left(\frac{-\sum\limits_{i=1}^{N} x_{1,i} x_{2,i}}{\sum\limits_{i=1}^{N} x_{1,i}^2 \sum\limits_{i=1}^{N} x_{2,i}^2 - (\sum\limits_{i=1}^{N} x_{1,i} x_{2,i})^2} \right),$$

$$S^2(b_0) = \frac{S^2(\hat{U})}{N} + \bar{X}_1^2 S^2(b_1) + \bar{X}_2^2 S^2(b_2).$$

The purpose of computing a multiple regression rather than a set of simple regressions, one for each explanatory variable, is to obtain the effect of an explanatory variable when other explanatory variables are held constant. The slope coefficient of a variable X_1 in a multiple regression is referred to as the *partial effect* of variable X_1. In the two-variable case, the slope coefficient b_1 is the effect of a unit increase in variable X_1 on variable Y, when variable X_2 is held constant.

Suppose the population regression equation is

$$Y_i = \beta_0 + \beta_1 X_{1,i} + \beta_2 X_{2,i} + U_i,$$

where the residual is normally distributed with a constant variance for all values of $X_{1,i}$ and $X_{2,i}$, not autocorrelated, and not correlated with $X_{1,i}$ or $X_{2,i}$. Then the coefficients of the estimated multiple regression are best, linear, unbiased estimates of the true population parameters. (See the introduction to Chapter 5.) The simple regression equation in the population could be written as

$$Y_i = \beta_0 + \beta_1 X_{1,i} + V_i,$$

where $V_i = \beta_2 X_{2,i} + U_i$. If variable X_2 has a partial effect on Y ($\beta_2 \neq 0$), and if X_1 and X_2 are correlated [$R(X_1, X_2) \neq 0$], then variables X_1 and V are correlated.†

† Using lower-case letters to designate deviations from means, $v_i = \beta_2 x_{2,i} + u_i$ and the covariance of x_1 and v is

$$\frac{\sum\limits_{i=1}^{N} x_{1,i} v_i}{N}.$$

Then

$$\sum_{i=1}^{N} x_{1,i} v_i = \sum_{i=1}^{N} x_{1,i}(\beta_2 x_{2,i} + u_i) = \beta_2 \sum_{i=1}^{N} x_{1,i} x_{2,i} + \sum_{i=1}^{N} x_{1,i} u_i = \beta_2 \sum_{i=1}^{N} x_{1,i} x_{2,i} \neq 0,$$

since X_1 and U are uncorrelated. Thus, X_1 and V are correlated.

TABLE 1

Direction of the Bias in the Algebraic Value of the Slope of an Explanatory Variable (X_1) Due to an Omitted Variable (X_2)†

Simple correlation of omitted variable with explanatory variable (X_1)	Partial effect of omitted variable on dependent variable (β_2)		
	Positive	Zero	Negative
Positive	Upward bias	No bias	Downward bias
Zero	No bias	No bias	No bias
Negative	Downward bias	No bias	Upward bias

† The two explanatory variable regression model is $Y_i = \beta_0 + \beta_1 X_{1,i} + \beta_2 X_{2,i} + U_i$.

If an explanatory variable and the residual are correlated, the estimated coefficient for the explanatory variable is a biased estimate of the true population parameter (see Table 1). A bias in a regression coefficient due to an omitted explanatory variable is referred to as a *specification bias*, since it arises from an imperfect specification of the relationships existing in the population.

Suppose, however, that variables X_1 and X_2 are uncorrelated $[R(X_1, X_2) = 0]$ in the population. Then the two estimates of the coefficient of X_1 obtained from computing the multiple regression and the simple regression would both be unbiased estimates of the population parameter. The coefficient obtained from the multiple regression would have a smaller standard error, and therefore a greater reliability, than the coefficient estimated from the simple regression.‡ In the multiple

‡ The standard error of a regression coefficient in a simple or multiple regression is proportional to the residual variance. The residual in the simple regression is

$$\hat{V}_i = b_2 X_{2,i} + \hat{U}_i.$$

The residual variance is

$$S^2(\hat{V}) = b_2{}^2 S^2(X_2) + S^2(\hat{U}),$$

since Cov $(X_2, \hat{U}) = 0$. Then $S^2(\hat{U})$ is smaller than $S^2(\hat{V})$ if $b_2 \neq 0$; that is, the residual variance is smaller in the two explanatory variable regression.

regression the explanatory variable X_2 is "accounting for" some of the previously unexplained residual variation.

Thus explanatory variables should be added to a regression model to reduce bias and to increase reliability. However, adding explanatory variables is not without cost. First, the intercorrelation among explanatory variables tends to increase as more variables are added. If the intercorrelation becomes sufficiently high, multicollinearity results in an increase in the standard errors of the regression parameters and therefore a decrease in their reliability. In the case of perfect multicollinearity in which one explanatory variable can be expressed as a linear function of one or more of the other explanatory variables, unique solutions cannot be computed for the regression coefficients.

Second, the addition of explanatory variables reduces the number of degrees of freedom (df). The number of degrees of freedom is the number of randomly selected observations minus the number of estimated regression coefficients. If the residual variance in the population is not known to be normally distributed, the larger the number of degrees of freedom, the closer is the approximation of the sampling distribution of the regression coefficients to the normal distribution. The distribution is approximately normal for $df > 30$. The t-ratio and F-ratio for testing hypotheses are strictly applicable only when the sampling distribution of the coefficients is normal.

Third, the collection and processing of data on additional variables is costly in terms of research time and computer time. Fourth, simpler models are always to be preferred to more complicated models since they are easier to interpret.

Data. Problems 1 through 9 use the research question developed for Chapter 5, with additional data on two new explanatory variables.

In a study of fertility a random sample of ten newly married couples were asked the number of children they desired to have (X), and whether the wife (W) and the husband (H) were college graduates. The education variables take the value of unity for college graduates and zero for those who are not college graduates. Twenty years later all ten couples were asked the number of children they actually had (Y). It is "known" that the residuals in the population are normally distributed. The results are presented in Table 2.

TABLE 2

Results of Study of Ten Randomly Selected Married Couples

Couple	Number of children		Education	
	Actual (Y)	Desired (X)	Wife (W)	Husband (H)
1	0	0	1	1
2	2	1	0	0
3	1	2	1	1
4	3	1	0	0
5	1	0	0	0
6	3	3	0	0
7	4	4	0	0
8	2	2	1	0
9	1	2	1	1
10	2	1	0	0

Problems

1. Two Explanatory Variable Model and the Effect of an Omitted Variable
2. Multicollinearity
3. Multiple Coefficient of Determination and Incremental Explanatory Power
4. Partial Correlation Coefficients
5. Beta Coefficients
6. Hypothesis Testing of Regression Coefficients—One Equation
7. Prediction from a Multiple Regression
8. Dichotomous (or Dummy) Explanatory and Dependent Variables and a Probability Dependent Variable
9. Interaction Variables
10. Testing the Equality of Regression Coefficients—Two Equations
11. Time-Series Analysis and Autocorrelated Residuals

Problem 1. Two Explanatory Variable Model and the
Effect of an Omitted Variable

Q: Regress the actual number of children (Y) on the desired number (X) and the wife's education (W). Test the significance of the slope coefficients. What is the effect on the slope coefficient of X when W is added to the equation?

A: The hypothesized equation in the population is

$$Y_i = \beta_0 + \beta_1 X_i + \beta_2 W_i + U_i,$$

and the sample equation is

$$Y_i = b_0 + b_1 X_i + b_2 W_i + \hat{U}_i,$$

or, in terms of deviations from means, $y_i = b_1 x_i + b_2 w_i + \hat{u}_i$. Since there are three coefficients to be estimated, there are three normal equations:

$$\sum_{i=1}^{N} x_i y_i = b_1 \sum_{i=1}^{N} x_i^2 + b_2 \sum_{i=1}^{N} x_i w_i,$$

$$\sum_{i=1}^{N} w_i y_i = b_1 \sum_{i=1}^{N} x_i w_i + b_2 \sum_{i=1}^{N} w_i^2,$$

$$\bar{Y} = b_0 + b_1 \bar{X} + b_2 \bar{W},$$

assuming that the residual is uncorrelated with X [$E(x\hat{u}) = 0$] and W [$E(w\hat{u}) = 0$] and that the residual has a zero mean [$E(\hat{u}) = 0$]. The residual variance is

$$S^2(\hat{U}) = \frac{\sum \hat{u}_i^2}{N - k},$$

where k is the number of estimated regression coefficients and N is the sample size. In this example $k = 3$.

$\hat{y}_i$ is the difference between the predicted value of the dependent variable for the ith observation and the mean: $\hat{y}_i = \hat{Y}_i - \bar{Y}$. Squaring both sides of the equation $y_i = \hat{y}_i + \hat{u}_i$ and summing across all N observations,

$$\sum_{i=1}^{N} y_i^2 = \sum_{i=1}^{N} \hat{y}_i^2 + \sum_{i=1}^{N} \hat{u}_i^2,$$

since $\sum \hat{y}\hat{u} = 0$.

Then

$$S^2(\hat{U}) = \frac{\displaystyle\sum_{i=1}^{N} y_i^2 - \sum_{i=1}^{N} \hat{y}_i^2}{N - k}.$$

The regression results are:

Dependent Variable: Y

Variable	Coefficient	Standard error	t-Value
Constant	1.453	0.309	4.71
X	0.628	0.136	4.63
W	-1.395	0.333	-4.20

$N = 10$, $\quad S^2(\hat{U}) = 0.264$, $\quad R^2 = 0.857$.

A t-value tests the hypothesis that the regression coefficient (whether a slope or the intercept) differs from zero. One degree of freedom is lost for each estimated regression coefficient; hence $df = N - k = 10 - 3 = 7$. Each coefficient is highly significant (the residual is assumed to be normally distributed). For $df = 7$, the critical t-ratio is 3.50 at a 1 percent level of significance.

The inclusion of W lowered the slope coefficient of X from 0.667 to 0.628. That is, excluding W from the equation in Problem 1 of Chapter 5 biased upward the slope coefficient of X. Returning to the first normal equation in the two explanatory variable model of this question, we have

$$\sum_{i=1}^{N} x_i y_i = b_1 \sum_{i=1}^{N} x_i{}^2 + b_2 \sum_{i=1}^{N} x_i w_i,$$

$$\frac{\displaystyle\sum_{i=1}^{N} x_i y_i}{\displaystyle\sum_{i=1}^{N} x_i{}^2} = b_1 + b_2 \frac{\displaystyle\sum_{i=1}^{N} x_i w_i}{\displaystyle\sum_{i=1}^{N} x_i{}^2}.$$

If we designate $b_1 = b_{yx.w}$ and $b_2 = b_{yw.x}$, we can write

$$b_{yx} = b_{yx.w} + b_{yw.x} b_{wx}.$$

If a variable (w) is correlated with an explanatory variable $(b_{wx} \neq 0)$ and has a nonzero partial effect on the dependent variable when the explanatory variable is held constant $(b_{yw.x} \neq 0)$, the exclusion of this variable from the regression equation biases the slope coefficient of the included explanatory variable.

In this example, b_{xw} is negative $(R_{xw} = -0.068)$, and from the regression, the partial effect of W on Y is negative $(b_2 = b_{yw.x} = -1.395)$. Since the product $b_{yw.x} \cdot b_{xw}$ is positive, b_{yx} is greater than $b_{yx.w}$. The slope coefficient of X is an upward biased estimate of the population parameter when the variable W is excluded from the regression (see Table 1).

Letting y be the dependent variable and Arabic numerals indicate independent variables, it can be shown that for three independent variables

$$b_{y1} = b_{y1.23} + b_{y2.13} \cdot b_{21} + b_{y3.12} \cdot b_{31}.$$

Problem 2. Multicollinearity

Q: (a) What is meant by "multicollinearity"? What are the consequences of high multicollinearity? Is there a test for multicollinearity? Are there procedures to correct for its adverse effects?

(b) Are the variables W and H highly correlated? What are the effects on the standard error of the coefficient of W and on the regression equation's explanatory power (R^2) when the variable H (husband's education) is added to the multiple regression of Y on X and W computed in Problem 1 of this chapter? The results from regressing Y on X, W, and H are:

Dependent variable: Y

Variable	Coefficient	Standard error	t-Value
Constant	1.508	0.271	5.57
X	0.595	0.120	4.97
W	−0.698	0.486	−1.44
H	−0.937	0.524	−1.79

$$S^2(\hat{U}_3) = 0.2011, \quad \sum_{i=1}^{N} \hat{U}_3{}^2 = 1.2066,$$

$$N = 10, \quad df = 6, \quad R^2 = 0.906.$$

A: (a) The situation that exists when one explanatory variable is a perfect linear function of one or more of the other explanatory variables in a multiple regression is referred to as *perfect multicollinearity*. When this occurs it is not possible to obtain a unique solution for the regression coefficients from the normal equations. (See the introduction to this chapter.)

If one explanatory variable is highly, but not perfectly, correlated with one or more of the other explanatory variables, the normal equations can be solved for unique estimates of the regression coefficients. The sample regression coefficients are unbiased estimates of the population parameters. However, multicollinearity biases upward the standard errors of the regression coefficients. This widens the confidence interval for the population parameter, or, in other words, reduces the reliability of the sample coefficient as an estimate of the population parameter. The addition of a highly collinear variable to a multiple regression can reduce the residual variance and therefore increase the regression model's explanatory power; that is, it can increase the reliability of the regression equation even though it decreases the reliability of some or all of the regression coefficients.

It should be asked, under what circumstances is multicollinearity a problem? Multicollinearity is undesirable if the object is to obtain reliable estimates of the coefficients of particular parameters or to test the statistical significance of particular variables. It is not a problem, however, if the object of the analysis is prediction *and* if it can be assumed that the same highly collinear structure that exists in the population from which the regression data are drawn (observations $1, \ldots,$ N) also exists in the population from which the data for prediction (observation $N + 1$) are selected.

There is no simple test for multicollinearity. One test procedure is to run a multiple regression of each explanatory variable on all the other explanatory variables. If there are h explanatory variables, compute h multiple regressions, each with $h - 1$ explanatory variables, and examine these multiple coefficients of determination (R^2). If none of these multiple coefficients of determination is "very high," multicollinearity is not a problem. There is no hard-and-fast rule for what size R^2 is "very high." The tolerable R^2 is generally higher the larger the size of the sample.†

A second procedure is to examine the sizes of the standard errors of the regression coefficients as additional explanatory variables are added to the multiple regression. The standard error of the coefficient of a variable X_1 will increase dramatically when one or more variables that are "highly" correlated with variable X_1 are added to the multiple regression.

There are several procedures for curing a multiple regression analysis in which some variables are highly correlated. These cures, however, are not without cost. First, one or more highly collinear explanatory variables can be deleted from the regression equation. This results in the loss of some of the regression equation's explanatory power. Second, *a priori* analysis or independent empirical evidence can be used to place additional

---

† See the footnote on p. 181.

constraints on the regression coefficients.† Information on the constraints may be costly to obtain, and the quality (reliability) of the final regression depends on the extent to which the constraints are correctly specified. Third, for the same structure, multicollinearity tends to be a less serious problem in larger samples. However, data from larger samples are more costly to collect and analyze.

(b) In the sample data the correlation of H with W is high: $R_{hw} = 0.80$. For such a small sample ($N = 10$) this correlation may be sufficient to cause the problems associated with multicollinearity. That is, W and H may be so highly intercorrelated that it is not possible to determine the individual effect of each variable on Y. The inclusion of H in the regression increases the standard error of the coefficient of W from 0.333 to 0.486. Both H and W are not separately significant. The inclusion of

H decreases the residual sum of squares ($\sum_{i=1}^{N} \hat{u}_i^2$) and increases

the explanatory power ($R^2 = 1 - \sum_{i=1}^{N} \hat{u}_i^2 / \sum_{i=1}^{N} y_i^2$) of the regres-

sion model from 0.857 to 0.906.

Problem 3. Multiple Coefficient of Determination and
Incremental Explanatory Power

Q: Use the data from the regression of Y on the explanatory variables X, W, and H presented in Problem 2.

(a) Test the significance of the explanatory power of the equation.

(b) What is the increment in explanatory power when W and H are added to the original simple regression? Is the increment significant?

---

† For example, if it is known that X_2 is a perfect linear function of X_1, then a unique solution for the coefficients of the regression $Y = b_0 + b_1X_1 + b_2X_2 + \hat{U}$ cannot be obtained using the normal equations. However, if it can be assumed that $b_2 = 2b_1$, there are three independent equations (two normal equations and the constraint on the coefficients) and three unknowns, and unique solutions can be computed.

A: (a) The model's explanatory power is

$$R^2 = 1 - \frac{\sum\limits_{i=1}^{N} \hat{U}_i^2}{\sum\limits_{i=1}^{N} y_i^2} = 0.906.$$

The adjustment for degrees of freedom is

$$\bar{R}^2 = 1 - \frac{\dfrac{\sum\limits_{i=1}^{N} \hat{U}_i^2}{N - k}}{\dfrac{\sum\limits_{i=1}^{N} y_i^2}{N - 1}} = 1 - \left(\frac{N - 1}{N - k}\right)(1 - R^2),$$

where k is the number of estimated regression coefficients and N is the size of the sample. In this example, $k = 4$, and

$$\bar{R}^2 = 1 - \tfrac{9}{6}(1 - 0.906) = 0.860.$$

The multiple correlation coefficient ($R = \sqrt{R^2}$) is the correlation between the observed Y and predicted values of the dependent variable: $R_{y\hat{y}} = \sqrt{0.906} = 0.952$. To test the statistical significance of this value from zero, where $H_0: \rho = 0$, $H_a: \rho \neq 0$, we use

$$t = \frac{R - \rho}{S(R)} = \frac{R}{\sqrt{(1 - R^2)[(k - 1)/(N - k)]}},$$

where k is the number of regression parameters estimated ($k = 4$). Substituting,

$$t = \frac{0.952}{\sqrt{(1 - 0.906)3/6}} = 4.40.$$

The number of degrees of freedom is $N - k = 6$. Note that this is a one-tailed test since the correlation between the observed and the predicted values of the dependent variable can never be negative in the least-squares model. For a one-tailed test, 2.5 percent level of significance, $df = 6$, the critical value is $t = 1.94$. The null hypothesis that the regression equation's explanatory power does not differ from zero is rejected.

This test is the same as the F-test for the entire equation:†

$$F_{(k-1,N-k)} = \frac{\sum \hat{y}^2/(k - 1)}{\sum \hat{u}^2/(N - k)} = \left(\frac{N - k}{k - 1}\right)\left(\frac{\sum y^2 - \sum \hat{u}^2}{\sum \hat{u}^2}\right)$$

$$= \left(\frac{N - k}{k - 1}\right)\left(\frac{1 - (\sum \hat{u}^2/\sum y^2)}{\sum \hat{u}^2/\sum y^2}\right)$$

$$= \left(\frac{N - k}{k - 1}\right)\left(\frac{R^2}{1 - R^2}\right)$$

$$= \left(\frac{R}{\sqrt{(1 - R^2)[(k - 1)/(N - k)]}}\right)^2 = t_{N-k}^2.$$

(b) The model's explanatory power (not adjusted for degrees of freedom) increased from $R_1^2 = 0.496$ to $R_3^2 = 0.906$ when the two education variables were entered:

$$R_1^2 = 1 - \frac{\sum (\hat{U})_1^2}{\sum (y)^2} \quad \text{and} \quad R_3^2 = 1 - \frac{\sum (\hat{U})_3^2}{\sum (y)^2}.$$

The fraction of the variation in the dependent variable

† For simplicity, $\sum_{i=1}^{N}$ is replaced by $\sum$, and the subscript i notation is suppressed.

unexplained by X, but explained by W and H, is

$$
\begin{aligned}
R^2_{ywh.x} &= \frac{\text{increase in explained variation}}{\text{variation left after controlling for } X} = \frac{R_3^2 - R_1^2}{1 - R_1^2} \\[2ex]
&= \frac{\sum (\hat{U})_1^2 - \sum (\hat{U})_3^2}{\sum (\hat{U})_1^2} = 1 - \frac{\sum (\hat{U})_3^2}{\sum (\hat{U})_1^2},
\end{aligned}
$$

where the "explained variation" is the reduction in the residual variance:

$$
R^2_{ywh.x} = 1 - \frac{1.21}{6.50} = 0.814.
$$

Thus 81.4 percent of the variation in the dependent variable left unexplained by the variable X is explained by W and H.

Is this increased explanatory power significant? Using the Fisher F-test,†

$$
F_{(l, N-k)} = \frac{\text{increase in explained variation}}{\text{unexplained variation}} = \frac{[\sum (\hat{U})_1^2 - \sum (\hat{U})_3^2]/l}{\sum (\hat{U})_3^2/(N - k)},
$$

k is the number of regression coefficients, $\sum (\hat{U})_3^2/(N - k)$ is the residual variance after the additional variables are entered, l is the number of additional variables, and $\sum (\hat{U})_1^2 - \sum (\hat{U})_3^2$ is the reduction in the residual sum of squares due to adding $l = 3 - 1 = 2$ explanatory variables. The F-statistic has l, $N - k$ degrees of freedom.

† The F-test is developed in Chapter 4. See the introduction and Problems 3 and 4 in Chapter 4, and Chapter 5, Problems 1 and 4.

In this example $N = 10$, $k = 4$, and $l = 2$. Then from p. 149 and p. 189,

$$\sum (\hat{U})_1{}^2 = 6.50,$$

$$\sum (\hat{U})_3{}^2 = 1.21,$$

$$F_{(l,N-k)} = \frac{[\sum (\hat{U})_1{}^2 - \sum (\hat{U})_3{}^2]/l}{\sum (\hat{U})_3{}^2/(N-k)} = \frac{(6.50 - 1.21)/2}{1.21/6} = 13.2$$

This is a one-tailed test. For l, $N - k$ or 2, 6 degrees of freedom and a one percent level of significance, the critical F-value is 10.92. The null hypothesis that the two added variables W and H do not have a significant partial effect on the model's explanatory power is rejected.

Thus even though neither of the two variables W and H is significant at a 10 percent level, when combined they are significant at a 1 percent level. When explanatory variables are highly intercorrelated (multicollinearity), it is not uncommon for one or more or all of the variables to have insignificant slope coefficients, yet for the set of variables to be significant!

The F-ratio for the inclusion of a set of variables and the adjusted coefficients of determination before and after inclusion are related. From the definition of $\bar{R}^2$ we know that

$$\sum \hat{U}_1{}^2 = (1 - \bar{R}_1{}^2)\left(\frac{\sum y^2}{N-1}\right)(N - k + l),$$

and

$$\sum \hat{U}_3{}^2 = (1 - \bar{R}_3{}^2)\left(\frac{\sum y^2}{N-1}\right)(N - k),$$

where in this problem $k = 4$ and $l = 2$. Then

$$F = \frac{(\sum \hat{U}_1{}^2 - \sum \hat{U}_3{}^2)/l}{\sum \hat{U}_3{}^2/(N-k)} = \left(\frac{N-k}{l}\right)\left(\frac{\sum \hat{U}_1{}^2}{\sum \hat{U}_3{}^2}\right) - \left(\frac{N-k}{l}\right),$$

$$\boxed{F_{(l,N-k)} = \left(\frac{1 - \bar{R}_1{}^2}{1 - \bar{R}_3{}^2}\right)\left(\frac{N-k+l}{l}\right) - \left(\frac{N-k}{l}\right).}$$

If the regression program prints the adjusted coefficient of determination, this equation provides a simple way of computing F-ratios for the inclusion of sets of variables. Note that:

if $\bar{R}_1^2 = \bar{R}_3^2$, then $F = 1$;

if $\bar{R}_1^2 > \bar{R}_3^2$, then $F < 1$; and

if $\bar{R}_1^2 < \bar{R}_3^2$, then $F > 1$.

Thus if the inclusion of additional variables lowers the adjusted coefficient of determination, the F-ratio for the inclusion of these variables is less than unity, and the set of additional variables is insignificant. In this problem,

$$F = \left(\frac{1 - \bar{R}_1^2}{1 - \bar{R}_3^2}\right)\left(\frac{N - k + l}{l}\right) - \left(\frac{N - k}{l}\right)$$

$$= \left(\frac{1 - 0.433}{1 - 0.860}\right)\left(\frac{10 - 4 + 2}{.2}\right) - \left(\frac{10 - 4}{2}\right) = 13.2.$$

Problem 4. Partial Correlation Coefficients

Q: Using the regression of actual number of children (Y) on desired number (X), wife's education (W), and husband's education (H), compute the partial correlation coefficients between the dependent variable and the explanatory variables. (Use the regressions given in Problem 2.)

A: The *partial correlation coefficient* between the dependent variable and an independent variable is the correlation between these two variables when the other variables in the model are held constant. The square of the partial correlation between Y and H is $r_{YH.XW}^2$, the proportion of the variation in Y that is left unexplained by X and W but that is explained by H. The proportion of the variation in Y left unexplained by X and W is $1 - R_{Y.XW}^2 = 1 - 0.857 = 0.143$. The increment in explanation from adding H to the equation is $R_{Y.XWH}^2 - R_{Y.XW}^2 = 0.906 - 0.857 = 0.049$.

The definition of the *partial coefficient of determination* is

$$r_{YH.XW}^2 = \frac{\text{increment in the explanation of } Y}{\substack{\text{unexplained variation in } Y \text{ when} \\ X \text{ and } W \text{ are held constant}}}$$

$$= \frac{R_{Y.XWH}^2 - R_{Y.XW}^2}{1 - R_{Y.XW}^2}.$$

Here

$$r_{YH.XW}^2 = \frac{0.049}{0.143} = 0.34,$$

$$r_{YH.XW} = -0.59.$$

(The sign of the partial correlation coefficient is the same as the sign of the slope coefficient of the variable.)

There is a short-cut method for computing partial correlation coefficients from a regression. Recall that if the null hypothesis is $H_0 : \rho = 0$, for a simple correlation coefficient,

$$t = \frac{R}{\sqrt{(1 - R^2)/df}},$$

where df is the number of degrees of freedom and $\sqrt{(1 - R^2)/df}$ is the standard error of R. Then $t^2(1 - R^2) = R^2(df)$, or

$$R^2 = \frac{t^2}{t^2 + df}.$$

Thus

$$r_{YH.XW}^2 = \frac{t^2}{t^2 + df},$$

where t is the t-ratio for the slope coefficient of H.† Computing, we have

$$r_{YH.XW}{}^2 = \frac{t^2}{t^2 + df} = \frac{(-1.79)^2}{(-1.79)^2 + 6} = 0.34$$

$$(df = N - k = 10 - 4 = 6)$$

$$r_{YX.HW}{}^2 = \frac{t^2}{t^2 + df} = \frac{(4.97)^2}{(4.97)^2 + 6} = 0.80,$$

$$r_{YW.HX}{}^2 = \frac{t^2}{t^2 + df} = \frac{(-1.44)^2}{(-1.44)^2 + 6} = 0.26,$$

$$r_{YH.XW} = -0.59,$$

$$r_{YX.HW} = 0.90,$$

$$r_{YW.HX} = -0.51.$$

(The partial correlation coefficient takes the same sign as the slope coefficient or its t-ratio.)

Neither the partial correlation coefficients (r) nor the partial coefficients of determination (r^2) add up to unity or to the regression's multiple correlation coefficient (R^2). Also the product of the correlation coefficients is neither unity nor R^2. Thus although partial correlation coefficients can be used to rank the relative importance of explanatory variables (an ordinal measure), they do not provide a cardinal measure. All that can be said about the relative importance of the variables from examining the partial correlation coefficients is that for explaining Y, in the three explanatory variable model, X is more important than H, which is more important than W.

Problem 5. Beta Coefficients

Q: Compute the contribution of each of the explanatory variables to the explanation of the dependent variable in the regressions of Y

---

† $r_{YH.XW}$ computed from the two procedures may differ slightly due to rounding errors.

on X, W, and H, and of Y on X and W. (The regressions are given in Problems 1 and 2.)

A: There exists an additive measure of explanatory power called the *beta coefficient*. If the regression equation has $k - 1$ explanatory variables and is

$$Y = b_0 + \sum_{i=1}^{k-1} b_i X_i + \hat{U},$$

then

$$S^2(Y) = \sum_{i=1}^{k-1} b_i^2 S^2(X_i) + \sum_{\substack{i \neq j \\ i=1 \\ j=1}}^{k-1} b_i b_j [\text{Cov}\ (X_i, X_j)] + S^2(\hat{U}).$$

If $\text{Cov}\ (X_i, X_j) = 0$ for all i, j (when $i \neq j$), then

$$1 = \sum_{i=1}^{k-1} \left(\frac{b_i S(X_i)}{S(Y)} \right)^2 + \frac{S^2(\hat{U})}{S^2(Y)}.$$

Since $R^2 = 1 - S^2(\hat{U})/S^2(Y)$,

$$R^2 = \sum_{i=1}^{k-1} \left(\frac{b_i S(X_i)}{S(Y)} \right)^2.$$

If we let

$$b_i^* = \frac{b_i S(X_i)}{S(Y)},$$

then,

$$R^2 = \sum_{i=1}^{k-1} (b_i^*)^2,$$

and b_i^* is referred to as the beta coefficient of variable i.

The beta coefficient is a measure of the statistical contribution of a variable (X_i) to the equation's explanatory power, and it is a pure number; that is, it has no units. However, the quality of the beta coefficient's measure of contribution to explanatory power is lower the greater the intercorrelation among the explanatory variables. If Cov $(X_i, X_j) \neq 0$ for any i, j combination (excluding $i = j$), then in general

$$R^2 \neq \sum_{i=1}^{k-1} (b_i{}^*)^2,$$

where $b_i{}^* = b_i S(X_i) / S(Y)$.

For the problem under investigation, $R(X,W) = -0.07$, $R(X,H) = -0.15$, and $R(H,W) = 0.80$. The beta coefficient would not be a good measure in the three explanatory variable model $(X, W, \text{and } H)$, but it would be a good measure in the two explanatory variable model (X,W).

For the three explanatory variable equation, we have

$$b_x{}^* = \frac{b_x S(X)}{S(Y)} = \frac{(0.595)(1.265)}{(1.197)} = 0.629,$$

$$b_w{}^* = \frac{b_w S(W)}{S(Y)} = \frac{(-0.694)(0.516)}{(1.197)} = -0.301,$$

$$b_h{}^* = \frac{b_h S(H)}{S(Y)} = \frac{(-0.937)(0.483)}{(1.197)} = -0.378,$$

and

$$\sum_{i=1}^{3} (b_i{}^*)^2 = 0.629.$$

However, $R^2 = 0.906$. Variable X is more important than variables W and H for explaining variation in Y.

In the two explanatory variable regression,

$$b_x{}^* = \frac{b_x S(X)}{S(Y)} = \frac{(0.628)(1.265)}{(1.197)} = 0.664,$$

$$b_w{}^* = \frac{b_w S(W)}{S(Y)} = \frac{(-1.395)(0.516)}{(1.197)} = -0.601,$$

and

$$\sum_{i=1}^{2} (b_i{}^*)^2 = 0.802.$$

However, $R^2 = 0.857$. Variable X is somewhat more important than variable W for explaining variation in Y.

The smaller the intercorrelation among the explanatory variables, the closer is the sum of the squared beta coefficients to the coefficient of determination; that is, the closer is $\sum_{i=1}^{k-1} (b_i{}^*)^2$ to R^2.

Problem 6. Hypothesis Testing of Regression Coefficients—
One Equation

Q: (a) Test the hypothesis that, holding the desired number of children constant, the effects of the level of schooling of the husband and wife on the actual number of children do not differ. (The regression is given in Problem 2.)

(b) Test the hypothesis that the sum of the partial effects of the schooling of the husband and wife equals -1.

A: (a) The null and alternative hypotheses are $H_0 : \beta_2 = \beta_3$ and $H_a : \beta_2 \neq \beta_3$ in the equation $Y_i = \beta_0 + \beta_1 X_i + \beta_2 W_i + \beta_3 H_i + U_i$. The t-ratio is

$$t = \frac{(b_2 - b_3) - (\beta_2 - \beta_3)}{S(b_2 - b_3)}$$

and

$$S^2(b_2 - b_3) = S^2(b_2) + S^2(b_3) - 2 \operatorname{Cov}(b_2, b_3).$$

The covariance of the slope coefficients would equal zero if we were testing the slopes in two separate equations based on independent samples. Since this is not the case here, the co-variance is not necessarily zero. The covariance, obtained from the computer print-out of the regression analysis, is -0.204. Then

$$S^2(b_2 - b_3) = (0.486)^2 + (0.524)^2 - 2(-0.204) = 0.919,$$

so

$$t = \frac{(b_2 - b_3) - (\beta_2 - \beta_3)}{S(b_2 - b_3)} = \frac{-0.694 + 0.937}{\sqrt{0.919}} = 0.25.$$

The null hypothesis of no difference in the slope coefficients of variables W and H is not rejected.

(b) $H_0:\beta_2 + \beta_3 = -1$, $H_a:\beta_2 + \beta_3 \neq -1$. Here

$$S^2(b_2 + b_3) = S^2(b_2) + S^2(b_3) + 2\,\mathrm{Cov}\,(b_2,b_3),$$

and

$$t = \frac{(b_2 + b_3) - (\beta_2 + \beta_3)}{S(b_2 + b_3)}.$$

Solving, we get

$$S^2(b_2 + b_3) = (0.486)^2 + (0.524)^2 + 2(-0.204) = 0.102,$$

$$t = \frac{(-0.694 - 0.937) - (-1)}{\sqrt{0.102}} = \frac{-0.631}{0.32} = 1.97.$$

The null hypothesis that the sum of the slope coefficients of variables W and H equals -1 is not rejected at a 5 percent level of significance.

Problem 7. Prediction from a Multiple Regression

Q: (a) Compute the predicted number of children for a family in which neither of the spouses is a college graduate ($W = 0, H = 0$) and which at marriage reported that they wanted two children ($X = 2$).

(b) What is the standard error of the predicted value?

(c) Compute the standard error of the intercept. (Use the regression given in Problem 2.)

A: The theory of prediction from a regression equation is developed in Problem 7 of Chapter 5. A test of residuals from the multiple regression indicates that the hypothesis of homoscedastic residuals cannot be rejected.

(a) The prediction equation is

$$\hat{Y}_i = b_0 + b_1 X_i + b_2 W_i + b_3 H_i.$$

The predicted number is

$$\hat{Y}_i = 1.508 + (0.595)(2) + (-0.694)(0) + (-0.937)(0),$$
$$= 2.698.$$

(b) To compute the standard error of the predicted value $S(\hat{Y})$, we use

$$\hat{Y}_i = b_0 + b_1 X_i + b_2 W_i + b_3 H_i + \hat{U}_i$$
$$= \bar{Y} + b_1 x_i + b_2 w_i + b_3 h_i + \hat{u}_i,$$

where lower-case letters designate deviations from means. Then

$$
\begin{aligned}
S^2(\hat{Y}) &= S^2(\bar{Y}) + x_i^2 S^2(b_1) + w_i^2 S^2(b_2) + h_i^2 S^2(b_3) \\
&\quad + 2x_i w_i \text{ Cov } (b_1,b_2) + 2x_i h_i \text{ Cov } (b_1,b_3) \\
&\quad + 2w_i h_i \text{ Cov } (b_2,b_3) + S^2(\hat{U}).
\end{aligned}
$$

This variance is computed under the assumptions that the sampling distributions of $\bar{Y}$ and $\hat{U}$ are not correlated with each other or with the slope coefficients and the values of the independent variables are held fixed. The sampling variance of the mean level of Y for given values of X is

$$S^2(\bar{Y}) = \frac{S^2(\hat{U})}{N}.$$

Most computer programs for regression analysis will compute the covariance matrix for the slope coefficients. For this problem the covariance matrix is

	b_1	b_2
b_2	-0.0048	
b_3	$+0.0096$	-0.2043

$\bar{X} = 1.6, \quad \bar{W} = 0.4, \quad \bar{H} = 0.3,$

Hence

$$S^2(\hat{Y}) = \frac{0.2011}{10} + (2 - 1.6)^2(0.120)^2 + (0 - 0.4)^2(0.486)^2$$

$$+ (0 - 0.3)^2(0.524)^2 + 2(2 - 1.6)(0 - 0.4)$$

$$\times (-0.0048) + 2(2 - 1.6)(0 - 0.3)(0.0096)$$

$$+ 2(0 - 0.4)(0 - 0.3)(-0.2043) + 0.2011$$

$$= 0.236$$

The 95 percent prediction interval is

$$\hat{Y} \pm S(\hat{Y})t_{0.025} = 2.698 \pm (0.486)(2.45) = 1.5 \text{ to } 3.9.$$

(c) Many computer programs do not print out the standard error of the intercept. This is the same as the standard error of prediction of the mean value of the dependent variable when all the independent variables take the value of zero. There is no residual error for the standard error of a predicted mean value. Then

$$b_0 = \hat{Y} = b_0 + b_1(0) + b_2(0) + b_3(0)$$
$$= \bar{Y} + b_1(0 - \bar{X}) + b_2(0 - \bar{W}) + b_3(0 - \bar{H});$$

and

$$S^2(b_0) = S^2(\bar{Y}) + (\bar{X})^2 S^2(b_1) + \bar{W}^2 S^2(b_2) + \bar{H}^2 S^2(b_3)$$
$$+ 2\bar{X}\bar{W} \text{ Cov } (b_1,b_2) + 2\bar{X}\bar{H} \text{ Cov } (b_1,b_3)$$
$$+ 2\bar{W}\bar{H} \text{ Cov } (b_2,b_3),$$

where

$$S^2(\bar{Y}) = \frac{S^2(\hat{U})}{N}.$$

Then,

$S(b_0) = 0.271$ is the standard error of the intercept.

Problem 8. Dichotomous (or Dummy) Explanatory and Dependent
Variables and a Probability Dependent Variable

Q: (a) Set up a regression equation to test the effect of region (Northeast, Midwest, South, West) and sex on the continuous dependent variable (Y) when the data are for individuals.
(b) Can an ordinary least-squares regression be used if the dependent variable is a dichotomous variable?
(c) Can an ordinary least-squares regression be used if the dependent variable is a probability or a proportion?

A: (a) *Dummy variables* or *dichotomous variables* can be created to test
the effects of qualitative variables (e.g., region, sex, race,
existence of law, discontinuity in the data). These variables
can also be used to convert quantitative variables into qualitative variables (e.g., years of schooling into low, medium, and
high schooling levels). The latter use of dummy variables
facilitates the testing for nonlinearities in the relationships
among the dependent and explanatory variables.
The regression equation can be written as

$$Y = b_0 + b_1X_1 + b_2X_2 + b_3X_3 + b_4X_4 + \hat{U},$$

where

$X_1 = 1$ for a resident of the Northeast and $X_1 = 0$ otherwise;
$X_2 = 1$ for a resident of the Midwest and $X_2 = 0$ otherwise;
$X_3 = 1$ for a resident of the South and $X_3 = 0$ otherwise;
$X_4 = 1$ for a male and $X_4 = 0$ for a female.

The four variables are treated as regular explanatory variables.
Two categories are not explicitly entered through variables
taking the value of unity. These are "West" and "female." If

a variable $X_5 = 1$ for a resident of the West were added, all possible residences would be included and the multiple correlation between X_5 and the variables X_1, X_2, and X_3 would be unity. The result is perfect multicollinearity. Similarly, if a variable $X_6 = 1$ were added for females, since all the subjects are either male or female, X_6 and X_4 would be perfectly negatively correlated.

The effect on Y of a person with the given characteristic is:

Female, West	b_0
Male, West	$b_0 + b_4$
Male, South	$b_0 + b_3 + b_4$
Female, Northeast	$b_0 + b_1$
Etc.	

For a qualitative variable with C classifications, $C - 1$ explanatory variables are created.

(b) A dichotomous variable can be a dependent variable in an ordinary least-squares regression analysis. For example, we could use microdata (data on individuals) to analyze the effect of several explanatory variables on whether a person has ever been married, where $Y_i = 1$ if ever-married and $Y_i = 0$ if never-married.

The predicted value of the dichotomous dependent variable is the probability that a person with given values of the explanatory variables has the characteristic under study, in this case is currently or has been married. However, predicted probability values less than zero or greater than unity can be

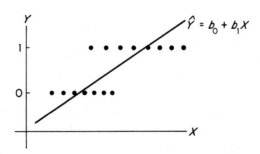

Fig. 2. Regression line for a dichotomous dependent variable.

obtained even though a probability has meaning only in the range 0.0 to 1.0 (see Figure 2). One solution is to define negative probabilities to be 0.0 and probabilities in excess of unity to be 1.0. A second solution is to use advanced techniques,† These techniques also correct for the heteroscedastic residuals that arise from an ordinary least-squares regression with a dichotomous dependent variable.

(c) In some analyses the dependent variable is a proportion or a probability (p) which, by definition, is bounded by zero and unity ($0 \leq p \leq 1$). A *probability dependent variable* generally arises from aggregating a dichotomous variable. For example, suppose we wish to study the determinants of school attendance. Whether a particular student i is in school on day j can be expressed as a dichotomous variable $Y_{ij} = 1$ if the student is in school, and $Y_{ij} = 0$ if the student is not in school. The proportion of days in a school year in which the student is in attendance is the mean value across school days j of the dichotomous variable Y_{ij}. If there are m school days, then

$$p_i = \frac{\sum\limits_{j=1}^{m} Y_{ij}}{m},$$

and p_i may be used as the dependent variable. Alternatively, the percentage of students in attendance in a particular time period (e.g., a day, a year) could be used as the dependent variable. If the time period is a day and there are N students, then

$$p_j = \frac{\sum\limits_{i=1}^{N} Y_{ij}}{N},$$

and p_j can be used as the dependent variable.

† See, for example, James Tobin, "Estimation of Relationships for Limited Dependent Variables," *Econometrica*, January, 1958, pp. 24–36, and Arthur S. Goldberger, *Econometric Theory* (Wiley, New York, 1964), pp. 248–254.

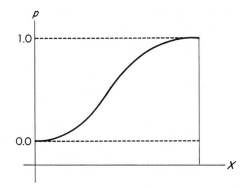

FIG. 3. Graphic representation of a logistic function, where p is a probability dependent variable. Logistic function:

$$p_i = 1/(1 + e^{-b_0 - b_1 X_{1,i} + \hat{U}_i}), \qquad Z_i = \ln[p_i/(1 - p_i)] = b_0 + b_i X_{1,i} + \hat{U}_i.$$

Since there are constraints on the values of a probability dependent variable ($0 \leq p \leq 1$), if p_i is regressed as a linear function of X_i, the standard errors may be biased due to heteroscedasticity. In addition, predicted values of the dependent variable $\hat{p}_i$ may fall outside the range of the interval 0.0 to 1.0. These problems can be prevented by converting the probability dependent variable into a continuous dependent variable by using a particular log linear regression equation called the *logistic function* (see Figure 3). If

$$\boxed{p_i = 1/(1 + e^{-b_0 - b_1 X_i - \hat{U}_i}),}$$

then†

$$Z_i = b_0 + b_1 X_i + \hat{U}_i,$$

† If we let $H_i = -b_0 - b_1 X_i - \hat{U}_i$, then $p_i = 1/(1 + e^H)$.

$$\frac{p_i}{1 - p_i} = \left(\frac{1}{1 + e^H}\right)\left(\frac{1}{1 - [1/(1 + e^H)]}\right) = \left(\frac{1}{1 + e^H}\right)\left(\frac{1 + e^H}{1 + e^H - 1}\right) = \frac{1}{e^H}.$$

Then

$$Z_i = \ln\left(\frac{p_i}{1 - p_i}\right) = \ln\left(\frac{1}{e^{H_i}}\right) = -H_i = b_0 + b_1 X_i + \hat{U}_i.$$

where $Z_i = \ln [p_i/(1 - p_i)]$, but $p_i \neq 1$ and $p_i \neq 0$.

Value of p_i	Value of $Z_i = \ln [p_i/(1 - p_i)]$
0.0	$-\infty$
0.5	0
1.0	$+\infty$

At the mean value of p_i,

$$\text{slope} = \frac{\Delta p}{\Delta X} = b_1(\bar{p} - \bar{p}^2),$$

$$\text{elasticity} = \frac{\Delta \ln p}{\Delta \ln X} = \frac{\Delta p}{\Delta X} \frac{\bar{X}}{\bar{p}} = b_1 \bar{X}(1 - \bar{p}).$$

Problem 9. Interaction Variables

Q: (a) Test the hypothesis that when the wife's education (W) and desired number of children (X) are the only explanatory variables, the deterrent effect of the wife's education does not depend on the desired number of children.

(b) Does the addition of the interaction term increase the model's explanatory power?

(c) Is *analysis of variance* (ANOVA) superior to regression analysis in testing for interaction?

A: (a) If the effect of W on Y depends on the level of X, there is an interaction between X and W. If a linear interaction is hypothesized, the regression equation is $Y_i = b_0 + b_1 X_i + b_2 W_i + b_3(X_i W_i) + \hat{U}_i$, where $(X_i W_i)$ is treated as if it were a new variable for the purposes of computing the regression coefficients. Then

$$\frac{\Delta Y}{\Delta W} = b_2 + b_3 X_i,$$

and the null hypothesis proposed in the question is rejected if b_3 differs from zero $(H_0: \beta_3 = 0)$.

The linear interaction is the simplest and most common form of interaction used in regression analysis. However, a large number of other forms of interactions (e.g., XW^2, X^2W, $X\sqrt{W}$, $X \div W$, etc.) could, in principle, be tested.

Computing the regression with the linear interaction term we have:

Dependent variable: Y

Variable	Coefficient	Standard error	t-Value
Constant	1.471	0.356	4.14
X	0.618	0.165	3.75
W	-1.471	0.659	-2.23
XW	0.049	0.360	0.14

$S^2(\hat{U}) = 0.3072$, $R^2 = 0.857$.

The slope coefficient of the interaction term does not differ significantly from zero. The null hypothesis that the effect of X does not depend on the level of W is tested using a linear interaction, and is not rejected.

(b) It was shown in Problem 3 of this chapter that the addition of explanatory variables to a regression increases the adjusted coefficient of determination ($\bar{R}^2$) only if the F-ratio for the added variables exceeds unity. If the "added variables" is only one variable, $F = t^2$. Hence $\bar{R}^2$ increases when one variable is added to the equation only if the absolute value of the variable's t-ratio exceeds unity ($|t| > 1.0$). The t-ratio of the linear interaction variable is $t = 0.14$, so the inclusion of this variable decreases the model's explanatory power adjusted for degrees of freedom ($\bar{R}^2$).

(c) *Analysis of variance* (ANOVA) is a multivariate statistical technique that can be used to test for any relationship among a set of two or more variables, including *any* form of interaction. ANOVA is limited, however, by the desirability of having the same sample size within each cell when there is a two-way (or k-way) classification of variables.

Regression analysis is not subject to this limitation on sample size. Regression analysis tests for a linear relationship between the dependent and explanatory variables. A continuous explanatory variable, however, can always be expressed as a set of dichotomous qualitative variables. For example, the continuous variable years of schooling (S) for a sample of individuals can be expressed as a set of dichotomous variables:

X_1 = 1 if 0 to 4 years of schooling; otherwise X_1 = 0.
X_2 = 1 if 5 to 8 years of schooling; otherwise X_2 = 0.
X_3 = 1 if 9 to 12 years of schooling; otherwise X_3 = 0.
X_4 = 1 if 13 to 16 years of schooling; otherwise X_4 = 0.
X_5 = 1 if more than 16 years of schooling; otherwise X_5 = 0.

(Only four of the five variables are entered in the regression.) This permits the use of regression analysis to test for "any relationship" between the dependent variable and an explanatory variable. Regression analysis can be used to test only for the forms of interaction specified by the investigator in the regression equation. In addition, regression analysis, but not ANOVA, can be used for prediction.

It is for these reasons that regression analysis now tends to be preferred to ANOVA, except when tests for "any interaction" between two explanatory variables are of primary importance.†

Problem 10. Testing the Equality of Regression Coefficients—
           Two Equations

Q: A regression analysis of income inequality (Y) among adult males used data from the southern and non-southern continental states of the United States and eight explanatory variables. The regression results are given in Table 3. Test the following hypotheses:

(a) The slope coefficient of explanatory variable X_1 is the same in the South and the non-South.
(b) The slope coefficients of explanatory variables X_1 and X_8 are the same in the non-South.
(c) The regression equations for the South and non-South are the same.

† For a not very mathematical exposition of ANOVA, see, for example, Helen M. Walker and Joseph Lev, *Statistical Inference* (Holt, Rhinehart and Winston, New York, 1953), Chapters 9 and 14.

TABLE 3
Results of a Study of Income Inequality (Y) among Adult Males[†]
dependent variable: Y; slope; t-ratios in parentheses

Variable	Non-South[‡]	South[‡]	All states[‡]
Intercept	−0.1444	−0.6207	−0.6788
X_1	3.1465 (10.70)	1.5344 (12.44)	1.8746 (10.61)
X_2	−0.0236 (−4.37)	0.0013 (0.39)	−0.0066 (−1.74)
X_3	0.0039 (1.58)	0.0098 (4.56)	0.0095 (5.22)
X_4	0.4859 (1.60)	1.2243 (4.31)	0.8627 (4.28)
X_5	0.0005 (0.12)	−0.0001 (−0.04)	−0.0005 (−1.62)
X_6	0.0009 (0.72)	−0.0001 (−0.13)	0.0002 (0.20)
X_7	0.0002 (0.87)	−0.0002 (−0.77)	−0.0000 (−0.19)
X_8	0.5872 (2.76)	0.1395 (2.94)	0.1330 (1.79)
No. of observations	32	17	49
R^2	0.9074	0.9882	0.9351
$S^2(Y)$	0.00578	0.00574	0.01323

[†]Data from B. R. Chiswick, "Racial Discrimination in the Labor Market: A Test of Alternative Hypotheses," *Journal of Political Economy*, November 1973, Appendix B.
[‡] The units of observation are the 48 continental states and the District of Columbia. The dependent variable is the variance of the natural log of income of adult males in each state.

A: (a) The test of the equality of two regression coefficients involves a t-test of differences, where the null hypothesis is equal slopes in the population:

$$H_0: \beta_{1,N} - \beta_{1,S} = 0, \qquad H_a: \beta_{1,N} - \beta_{1,S} \neq 0,$$

where N designates non-South and S designates South. Then

$$t_d = \frac{(b_{1,N} - b_{1,S}) - (\beta_{1,N} - \beta_{1,S})}{S(b_{1,N} - b_{1,S})},$$

$$S^2(b_{1,N} - b_{1,S}) = S^2(b_{1,N}) + S^2(b_{1,S}) - 2\,\text{Cov}\,(b_{1,N}, b_{1,S}).$$

Since the non-southern and southern samples were drawn independently of each other, $b_{1,N}$ is independent of $b_{1,S}$ and

Cov $(b_{1,N}, b_{1,S}) = 0$. Then

$$S(b_{1,N} - b_{1,S}) = \sqrt{S^2(b_{1,N}) + S^2(b_{1,S})},$$

$$S^2(b_{1,N}) = \left(\frac{b_{1,N}}{t_{1,N}}\right)^2 = \left(\frac{3.1465}{10.70}\right)^2 = 0.0865,$$

$$S^2(b_{1,S}) = \left(\frac{b_{1,S}}{t_{1,S}}\right)^2 = \left(\frac{1.5344}{12.44}\right)^2 = 0.0152,$$

$$S(b_{1,N} - b_{1,S}) = \sqrt{0.0865 + 0.0152} = 0.3189,$$

$$t_d = \frac{(3.1465 - 1.5344) - 0}{0.3189} = 5.05.$$

The number of degrees of freedom for the observed t-ratio is the sum of the degrees of freedom in the two separate samples:

$$df = df_N + df_S = (32 - 9) + (17 - 9) = 31.$$

For 31 degrees of freedom, and a 1 percent two-tailed alpha or type I error, critical $t = 2.75$. The null hypothesis of an equal partial effect of the variable X_1 on income inequality (Y) in the South and the non-South is rejected.

(b) The slope and intercept (constant) coefficients within a single regression equation are *not* statistically independent of each other. This introduces one important difference from the procedure used in the answer to part (a) of this problem. The null and alternative hypotheses are

$$H_0: \beta_1 - \beta_8 = 0, \qquad H_a: \beta_1 - \beta_8 \neq 0.$$

Then

$$t_d = \frac{(b_1 - b_8) - (\beta_1 - \beta_8)}{S(b_1 - b_8)},$$

$$S^2(b_1 - b_8) = S^2(b_1) + S^2(b_8) - 2\,\text{Cov}\,(b_1, b_8).$$

The covariance term, Cov (b_1, b_8), is not necessarily zero. Most regression programs have an option for computing the variance–covariance matrix of the regression slope coefficients. For this problem,

Cov (b_1, b_8) = 0.0004.

Then

$$S^2(b_1) = \left(\frac{b_1}{t_1}\right)^2 = \left(\frac{3.1465}{10.70}\right)^2 = 0.0865,$$

$$S^2(b_8) = \left(\frac{b_8}{t_8}\right)^2 = \left(\frac{0.5872}{2.76}\right)^2 = 0.0453,$$

$$S(b_1 - b_8) = \sqrt{0.0865 + 0.0453 - 2(0.0004)} = 0.3619,$$

$$t_d = \frac{(3.1465 - 0.5872) - 0}{0.3619} = 7.07.$$

The number of degrees of freedom for the t-test is the number of degrees of freedom in the regression equation: $df = N - k = 32 - 9 = 23$.

The null hypothesis of equal partial effects of variables X_1 and X_8 in the non-South is rejected.

(c) There are differences in the intercept and slope coefficients for the two regional regressions. We wish to test whether these differences are sufficiently large for us to reject the null hypothesis that the two *structures* are identical. If the two structures are identical and there is no sampling error, the sum of the squared residuals from the two separate regressions

$$\left(\sum_{i=1}^{N_1} U_1^2 + \sum_{i=1}^{N_2} U_2^2\right)$$

equals the sum of the squared residuals from the pooled sample

$$\left(\sum_{i=1}^{N_p} U_p^2\right), \text{ where } N_p = N_1 + N_2 \text{ is the number of observations.}$$

If the two population structures are identical but there is

sampling variability,

$$\sum_{i=1}^{N_1} U_1{}^2 + \sum_{i=1}^{N_2} U_2{}^2 \leq \sum_{i=1}^{N_p} U_p{}^2.$$

(Since twice as many parameters are estimated to obtain the sum of the residual sums of squares, this sum will generally be less than the pooled sum of squared residuals.) If the two structures differ, the pooled regression will perform far more poorly (larger residual sum of squares) than the two separate regressions. Thus, if $\sum_{i=1}^{N_1} U_1{}^2 + \sum_{i=1}^{N_2} U_2{}^2$ is sufficiently smaller than $\sum_{i=1}^{N_p} U_p{}^2$, the null hypothesis that the structures are the same is rejected.

The test of the null hypothesis of equal structures is called a *Chow test*; it uses the F-distribution:

$$F(k, N_p - 2k) = \frac{[(\sum_{i=1}^{N_p} U_p{}^2) - (\sum_{i=1}^{N_1} U_1{}^2 + \sum_{i=1}^{N_2} U_2{}^2)]/k}{(\sum_{i=1}^{N_1} U_1{}^2 + \sum_{i=1}^{N_2} U_2{}^2)/(N_p - 2k)},$$

where k is the number of parameters estimated in each regression, $N_p = N_1 + N_2$, and $(k, N_p - 2k)$ is the number of degrees of freedom of the F-statistic. The null hypothesis of the same (equal) population structures for the data in samples 1 and 2 is rejected if the observed F-ratio exceeds the critical F-ratio.

Using the data in this problem, we have

$$R^2 = 1 - \frac{\sum U^2}{\sum Y^2}, \qquad \sum U^2 = (\sum Y^2)(1 - R^2).$$

For the non-South, the sum of the squared residuals is

$$\sum_{i=1}^{N_1} U_1^2 = (0.00578)(32)(1 - 0.9074) = 0.01713.$$

For the South it is

$$\sum_{i=1}^{N_2} U_2^2 = (0.00574)(17)(1 - 0.9882) = 0.00115.$$

The sum is

$$\sum_{i=1}^{N_1} U_1^2 + \sum_{i=1}^{N_2} U_2^2 = 0.01713 + 0.00115 = 0.01828.$$

For the pooled regression (South and non-South data) with $N_p = N_1 + N_2$ observations, the sum of the squared residuals is

$$\sum_{i=1}^{N_p} U_p^2 = (0.01323)(49)(1 - 0.9351) = 0.04207.$$

Hence

$$F(9, 49 - 18) = \frac{(0.04207 - 0.01828)/9}{0.01828/(49 - 18)},$$

$$F(9, 31) = \frac{0.0026433}{0.0005896} = 4.48.$$

The critical F-value (one-tailed test) for $\alpha = 0.01$ is

$$F(9, 31) = 3.03.$$

The null hypothesis that the structures are the same is rejected.

Problem 11. Time-Series Analysis and Autocorrelated Residuals

Q: Table 4 presents time-series data on the income inequality of adult males from 1950 to 1968. Income inequality is measured by the variance of the natural log of income.

TABLE 4

Time-series Analysis of Income Inequality†

Year	Time (T)	Income inequality (Y)	Inequality of weeks worked (W)
1950	1	0.6341	0.1762
1951	2	0.5570	0.1307
1952	3	0.5295	0.1268
1953	4	0.5844	0.1233
1954	5	0.6545	0.1576
1955	6	0.6387	0.1256
1956	7	0.6312	0.1287
1957	8	0.6434	0.1583
1958	9	0.6447	0.1910
1959	10	0.6483	0.1499
1960	11	0.6635	0.1599
1961	12	0.6858	0.1737
1962	13	0.6413	0.1743
1963	14	0.6318	0.1569
1964	15	0.6307	0.1422
1965	16	0.6282	0.1321
1966	17	0.5808	0.1207
1967	18	0.5675	0.1112
1968	19	0.5609	0.1113

† For the source of the data and the logic behind the form of the variables, see Barry R. Chiswick and Jacob Mincer, "Time Series Changes in Income Inequality Since 1939, with Projections to 1985," *Journal of Political Economy, Supplement*, May/June 1972, s34–s66. The data are for adult males 25 to 64 years of age. $Y = \sigma_2(\ln \text{income})$ and $W = \sigma^2(\ln \text{weeks worked})$.

(a) Regress income inequality on a nonlinear time trend to test the hypothesis that income inequality has changed over time.

(b) What is meant by autocorrelated residuals? Why are autocorrelated residuals a problem?

(c) Test the regressions computed in part (a) for autocorrelated residuals.

(d) Correct for autocorrelated residuals and test for a time trend in income inequality.

A: (a) The regression of income inequality on time and time squared is:

Dependent Variable: Y

Variable	Regression coefficient	Standard error	t-Ratio
Constant	0.5395	0.0234	23.07
Time (T)	0.0223	0.0054	4.14
Time squared (T^2)	−0.0011	0.0003	−4.25

$N = 19, \quad \sum_{i=1}^{N} \hat{U}_i^2 = 0.0149, \quad R^2 = 0.53.$

The regression equation indicates a significant curvilinear time trend. Inequality increased from 1950 to 1959 and then declined. $[\partial Y / \partial T = 0.0223 - 2(0.0011)T = 0, \quad T = 0.0223/0.0022 = 10$ at the maximum.]

To test whether these trends are a consequence of cyclical variations in employment, we can add a variable for the relative inequality of weeks worked $[W = \sigma^2(\ln \text{weeks worked})]$ to the regression equation. The relative inequality of weeks worked increases in a recession and decreases in an expansion.

Dependent variable: Y

Variable	Regression coefficient	Standard error	t-Ratio
Constant	0.4276	0.0409	10.47
Time	0.0152	0.0049	3.08
Time squared	−0.0007	0.0002	−2.86
Inequality of weeks worked (W)	0.9074	0.2936	3.09

$N = 19, \quad \sum_{i=1}^{N} \hat{U}_i^2 = 0.0091, \quad R^2 = 0.7129.$

Assuming the residuals are normally distributed, for $df = 19 - 4 = 15$, $\alpha = 0.05$, the critical $t = 2.13$. The time and weeks worked variables are statistically significant. Inequality reached a peak in 1960. $[\partial Y/\partial T = 0.0152 - 2(0.0007)T, \ T = 0.0152/0.0014 \approx 11.]$

Note: The fallacy of predicting the dependent variable for values outside the range of the explanatory variable is easily seen using these regression equations. Sufficiently far in the past and in the future, the regression predicts negative income inequality!

(b) One problem in time-series data is *serially correlated (autocorrelated) residuals*. If the residuals are positively correlated, a positive (negative) residual last year suggests that a positive (negative) residual is likely this year. Positive autocorrelation is common because the effects of random occurrences and omitted variables tend to persist over a period of time. Negative autocorrelation is less common and is usually a consequence of an omitted variable which takes alternating values (winter vs. summer, day vs. night, etc.). Positively and negatively correlated residuals are illustrated in Figure 4.

Positive autocorrelation biases downward the estimate of the residual variance. This in turn biases downward the standard error of the intercept and regression slope coefficients, and biases upward the model's explanatory power (R^2). Downward biased standard errors may result in the rejection of the null hypothesis of no statistically significant effect of an explanatory variable when, in fact, there is no significant effect. Thus we should test the regression equations in part (a) to see whether our conclusions were influenced by autocorrelated residuals.

(c) Although inspection of the residuals is often used to "test" for autocorrelation, the most common statistical procedure is the Durbin-Watson test.

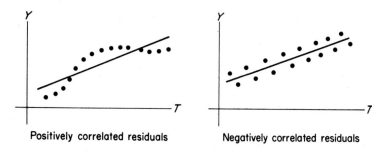

Positively correlated residuals   Negatively correlated residuals

FIG. 4

The *Durbin-Watson statistic* is

$$d = \frac{\sum_{t=2}^{N} (\hat{U}_t - \hat{U}_{t-1})^2}{\sum_{t=1}^{N} \hat{U}_t^2},$$

where $\hat{U}_t$ is the residual in time period t. The number of differences in the numerator is equal to the number of successive residuals for which there are no gaps (missing observations). In this problem, all years between 1950 and 1968 are in the data series, so there are $N - 1 = 18$ differences in the numerator.

Correlation of error terms $(\hat{U}_t, \hat{U}_{t-1})$	Range of d†
Positive	0–2
Zero	2
Negative	2–4

$$† \; d = \frac{\sum_{t=2}^{N} (\hat{U}_t - \hat{U}_{t-1})^2}{\sum_{t=1}^{N} \hat{U}_t^2} = \frac{\sum_{t=2}^{N} \hat{U}_t^2 - 2\sum_{t=2}^{N} \hat{U}_t\hat{U}_{t-1} + \sum_{t=2}^{N} \hat{U}_{t-1}^2}{\sum_{t=1}^{N} \hat{U}_t^2}. \text{ For large samples, and}$$

homoscedastic residuals, $\sum_{t=2}^{N} \hat{U}_t^2 \simeq \sum_{t=2}^{N} \hat{U}_{t-1}^2.$

$$d = \frac{\sum_{t=2}^{N} (\hat{U}_t - \hat{U}_{t-1})^2}{\sum_{t=1}^{N} \hat{U}_t^2} \simeq 2\left(\frac{\sum_{t=2}^{N} \hat{U}_t^2 - \sum_{t=2}^{N} \hat{U}_t\hat{U}_{t-1}}{\sum_{t=1}^{N} \hat{U}_t^2} \right) \simeq 2\left(1 - \frac{\sum_{t=2}^{N} \hat{U}_t\hat{U}_{t-1}}{\sum_{t=1}^{N} \hat{U}_t^2} \right).$$

Thus $d \simeq 2(1 - r)$, where r is the simple correlation coefficient between successive residuals (first-order linear autocorrelation coefficient).

The Durbin-Watson statistic does not have unambiguous critical values (see Figure 5). The values of d and the conclusions are:

Values of d	Conclusion
0 to d_l	Accept hypothesis of positive autocorrelation
d_l to d_u	Neither accept nor reject null hypothesis
d_u to 2, 2 to $4d_u$	Accept null hypothesis of no autocorrelation
$4d_u$ to $4d_l$	Neither accept nor reject null hypothesis
$4d_l$ to 4	Accept hypothesis of negative autocorrelation

The parameters of the d-statistic are k, the number of regression coefficients, and N, the number of observations. The values of d_l and d_u are smaller, the larger is k and the smaller is N. See Table 4 in the Appendix.

Regression of Y on	Observed d	k	N	Critical d ($\alpha = 0.05$) d_l	d_u
T, T^2	1.41	3	19	0.96	1.41
T, T^2, W	1.48	4	19	0.86	1.55

When income inequality is regressed on the nonlinear time trend, the Durbin-Watson statistic is on the borderline of

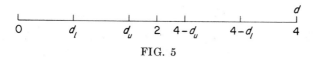

FIG. 5

accepting the hypothesis of no autocorrelation. When the inequality of weeks worked is added to the regression equation, we can neither accept nor reject the hypothesis of no autocorrelation.

(d) There are several techniques for correcting for autocorrelated residuals.

One procedure is to add to the regression equation one or more explanatory variables that are the cause of the autocorrelation. For example, if we had monthly data on income inequality and if inequality is, say, lower in the summer months than in the winter months, a dummy variable for season or a set of dummy variables for months might eliminate the autocorrelation.

A second procedure is to use *first differences*. If

$$Y_{t,i} = b_0 + b_1 X_{t,i} + \hat{U}_{t,i} \quad \text{and} \quad Y_{t-1,i} = b_0 + b_1 X_{t-1,i} + \hat{U}_{t-1,i},$$

then

$$Y_{t,i} - Y_{t-1,i} = b_1(X_{t,i} - X_{t-1,i}) + \hat{V}_{t,i},$$

where $\hat{V}_{t,i} = \hat{U}_{t,i} - \hat{U}_{t-1,i}$.

Regressing the change in the dependent variable on the change in the explanatory variables, we obtain estimates of the regression coefficients and test for autocorrelation.

If

$$Y_i = b_0 + b_1 T_i + b_2 T_i^2 + b_3 W_i + \hat{U}_i,$$

then†

$$\Delta Y_i = b_1 + (2b_2) T_i + b_3(\Delta W_i) + \hat{V}_i.$$

$\dagger \dfrac{dY}{dT} = b_1 + (2b_2)T + b_3\left(\dfrac{dW}{dT}\right).$

Dependent variable: ΔY

Variable	Coefficient	Standard error	t-Ratio
Constant	0.0045	0.0187	0.24
T	-0.0008	0.0016	-0.51

$$N = 18, \quad \sum_{i=1}^{N} \hat{V}_i^2 = 0.020, \quad R^2 = 0.016, \quad d = 1.34.$$

Dependent variable: ΔY

Variable	Coefficient	Standard error	t-Ratio
Constant	0.0072	0.0158	0.46
T	-0.0008	0.0013	-0.58
ΔW	0.8621	0.3149	2.74

$$N = 18, \quad \sum_{i=1}^{N} \hat{V}_i^2 = 0.013, \quad R^2 = 0.34, \quad d = 1.65.$$

The critical values of d for $\alpha = 0.05$, and $N = 18$ (18 differences) are:

k	d_l	d_u
2	1.03	1.26
3	0.93	1.40

(See Table 4 in the Appendix.)

For both regressions, the procedure of first differences results in the acceptance of the null hypothesis of zero autocorrelation. In the first-difference equation the insignificant constant term (b_1) and the insignificant slope of T ($2b_2$) mean that we can reject the hypothesis of a time trend for income inequality in the period under study. There is, however, a significant effect of the inequality of weeks worked on income inequality.

First differences is a special case of a third technique—the first-order autoregressive procedure.

If

$$Y_t = b_0 + b_1 X_t + \hat{U}_t \quad \text{and} \quad Y_{t-1} = b_0 + b_1 X_{t-1} + \hat{U}_{t-1},$$

then

$$Y_t - rY_{t-1} = b_0(1 - r) + b_1(X_t - rX_{t-1}) + \hat{V}_t,$$

where $\hat{V}_t = \hat{U}_t - r\hat{U}_{t-1}$ and r is the correlation coefficient of U_t and U_{t-1}. Recall from the footnote on p. 220 that $d \simeq 2(1 - r)$. Then the regression

$$Y_t^* = b_0^* + b_1^*X_t^* + \hat{V}_t^*,$$

where

$$Y_t^* = Y_t - rY_{t-1}, \qquad X_t^* = X_t - rX_{t-1},$$

$$b_0^* = b_0(1 - r), \text{ and } b_1^* = b_1,$$

permits the computation of b_0 and b_1 if a value is specified for r. If $r = 1.0$, the first-order autoregressive procedure is the same as the procedure of first differences.

Chapter 7

Simultaneous Systems of Equations

Chapters 5 and 6 on simple and multiple regression analysis present techniques for estimating the parameters of a model that has only one *structual (behavioral) equation.* A structural equation relates a dependent variable to one or more explanatory variables and a residual. Many studies in economics, business, and the other social sciences use models in which there is more than one structural equation. For example, the analysis of the determinants of the price and quantity of a particular commodity requires at least two structural equations, the supply equation and the demand equation for the commodity.

Suppose there is a system of S equations with S dependent variables. If the value of each of the explanatory variables in the system of equations is determined independently of the S dependent variables, the set of equations can be estimated by computing S single equation ordinary least-squares (OLS) regressions. The system of equations is referred to as a *simultaneous system* if one or more of the S dependent variables are also explanatory variables in one or more of the equations. The dependent variables in a simultaneous system are referred to as *endogenous variables.* The variables in a simultaneous system that are determined by factors outside the model are referred to as *exogenous variables.*

The parameters of the structural equations in a simultaneous system of equations should not be estimated by ordinary least squares. If an explanatory variable X is determined by the variable Y, the slope coefficient of the ordinary least-squares regression of Y on X is biased.†
This is referred to as a *simultaneous equations bias.* Two procedures considered here for obtaining estimates of the parameters of the structural equations in a simultaneous system are *indirect least squares* (ILS)

---

† For a mathematical proof see a theoretical econometrics textbook, such as Teh-wei Hu, *Econometrics: An Introductory Analysis* (University Park Press, Baltimore, 1973), pp. 122–125.

and *two-stage least squares* (2SLS). More advanced techniques have been developed, but they are beyond the scope of this book.

Indirect least squares is the computationally simpler of the two procedures. However, ILS provides unique estimates of the coefficients of a system of structural equations only if the equations are exactly identified. 2SLS can be used to estimate the structure of an exactly identified or an over-identified system of equations. The two procedures result in identical estimates of the structural parameters for an exactly identified system of equations. Since most econometric studies involve over-identified structural equations, 2SLS is the more widely used procedure. Moreover, 2SLS provides direct information about the significance of explanatory variables. Computer programs with 2SLS options make the procedure relatively easy to use. Three characteristics of ILS, 2SLS, and OLS estimates of structural equations are summarized in Table 1.

TABLE 1

Characteristics of Alternative Procedures for Estimating the Coefficients of a Simultaneous System of Equations

	Ordinary least squares	Indirect least squares	Two-stage least squares
Identification	Identification is not required	Exactly identified equations†	Exactly identified or over-identified equations†
Structural coefficients	Biased	Consistent estimator, biased for small samples.‡	Consistent estimator, biased for small samples.‡
Efficiency	Minimum residual variance	Not minimum residual variance	Not minimum residual variance

† The estimated structural coefficients are identical for exactly identified systems of equations.
‡ For a consistent estimator the bias decreases to zero as the sample size becomes very large (approaches infinity).

Problems

1. Simultaneous Equations Bias
2. Identification
3. Indirect Least Squares (ILS)
4. Two-Stage Least Squares (2SLS)
5. Instrumental Variables

Problem 1. Simultaneous Equations Bias

Q: Data are available on the price (P) and quantity (Q) sold per unit of time for a particular commodity, as well as for other variables that influence the supply and demand curves for this commodity. Is a single-equation multiple regression of quantity (Q) on price (P) and other variables a demand equation, a supply equation, or a hybrid? What is meant by a "simultaneous equations bias"?

A: Variables in a simultaneously determined system of equations are classified as either exogenous or endogenous. An *exogenous variable* in a system influences (determines) the value of some other variables in the system but is itself determined entirely by factors outside the model. For example, for a time-series study of the supply and demand for corn, the amount of rainfall is an exogenous variable, if we assume no successful efforts are made to artificially induce rain by cloud seeding. An *endogenous variable* in a system of equations influences (determines) the value of other variables in the system and is itself influenced (determined) by the variables in the model. In a time-series study of the supply and demand for corn, the price of corn both determines the quantity of corn demanded and is itself determined by the quantity of corn produced.

A supply and demand model can, for example, be written as:

Demand: $Q_i = a_0 + a_1 P_i + a_2 X_i + \hat{U}_i,$
Supply: $P_i = b_0 + b_1 Q_i + b_2 Z_i + \hat{V}_i,$

where price (P) and quantity (Q) are endogenous variables, X and Z represent the set of one or more exogenous variables in each equation, and U and V are the residuals. There is always one *behavioral* (or *structural*) *equation* for each endogenous variable in a system of equations. Some, but not all, of the exogenous variables can appear in both equations. (This system of equations can be written with two identities: supply price equals demand price, $P_S = P_D$, and supply quantity equals demand quantity, $Q_S = Q_D$.)

The observations on price (P) and quantity (Q) are the equilibrium points. Suppose neither the demand curve nor the supply curve is stable across the observations. The single-equation multiple regression of quantity on price and the exogenous variables in the demand equation does *not* result in the computation of the demand equation. The computed slope coefficient of the price variable is a biased estimate of the slope coefficient of price in the true demand equation. It is biased because the residual (U) is in general correlated with the variable price (P). The residual (U) is the deviation of the observed quantity (Q) from the predicted value of quantity holding price constant $(\hat{Q})$. The observed price (P) is, however, itself a function of the observed quantity. The pure effect of P on Q is clouded by the effect of Q on P. The result is a residual that is not necessarily uncorrelated with price, and the slope coefficient of the variable price is biased. This is referred to as a *simultaneous equations bias*.

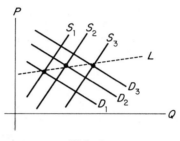

FIG. 1

Figure 1 illustrates shifting supply and demand curves. Here line L is the ordinary least-squares line obtained from regressing quantity on price. The slope of L is clearly a biased estimate of the slope of the demand curve.

Similarly, if price is regressed on quantity and the exogenous variables in the supply equation, the result is *not* a supply equation. Price and quantity are interdependent in a one-equation regression; quantity (Q) and the residual (V) are in general correlated and the slope coefficient of quantity is biased (simultaneous equations bias).

A single-equation multiple regression of quantity on price and a set of exogenous variables is neither a supply equation nor a demand equation, but rather is a hybrid. Ordinary least-squares single-equation multiple regression techniques cannot be used to estimate the structural (behavioral) equations in a system of simultaneous equations.

Problem 2. Identification

Q: What is meant by the "identification" of the parameters of a structural equation in a system of simultaneous equations? Under what circumstances is it possible to identify the structural parameters of a two-equation supply and demand model?

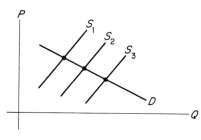

FIG. 2

A: Identification refers to the problem of computing the parameters of the structural equations.

Suppose it is known that the demand curve for a commodity is perfectly stable but the supply curve differs across observations, as shown in Figure 2. The observations on price and quantity trace out the demand curve. The regression of quantity on price is a demand equation since there is no simultaneous equations bias. It is not possible, however, to estimate the parameters of the supply equation. Thus a stable demand curve and a shiftng supply curve permit the identification of the parameters of the demand equation, but not of the supply equation.

If it is known that the supply curve is perfectly stable but the demand curve varies, the observations on price and quantity trace out the supply curve (see Figure 3). The supply curve is identified but the demand curve is under-identified.

The identification of the demand curve does not require that it be perfectly stable, but only that the supply curve contain one or more exogenous variables that shift the supply curve but do not affect the demand curve. If there is only one such explanatory

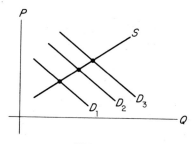

FIG. 3

variable in the supply equation, the demand equation is *exactly identified*. If there are two or more such variables, the demand equation is *over-identified*. The identification of the supply equation requires that the demand equation contain one or more exogenous explanatory variables that do not affect the supply curve. Table 2 indicates the relationship between alternative structures of the two equations in a supply and demand model and statistical identification.

TABLE 2
Structural Demand and Supply Equations and Identification†

Equation	Structure	Identification
Demand	$Q = f(P, X_1, X_2)$	Under-identified
Supply	$P = g(Q, X_1, X_2)$	Under-identified
Demand	$Q = f(P, X_1)$	Exactly identified
Supply	$P = g(Q, X_2)$	Exactly identified
Demand	$Q = f(P, X_1, X_2)$	Over-identified
Supply	$P = g(Q, X_3, X_4)$	Over-identified
Demand	$Q = f(P, X_1, X_2)$	Under-identified
Supply	$P = g(Q, X_1)$	Exactly identified
Demand	$Q = f(P, X_1, X_2, X_3)$	Exactly identified
Supply	$P = g(Q, X_4)$	Over-identified

† P = price, Q = quantity, X_i for $i = 1, \cdots, 4$ = exogenous variable i.

Problem 3. Indirect Least Squares

Q: Use indirect least squares to compute the parameters for each of the following two-equation supply and demand systems. [P and Q are price and quantity and X_1, X_2, and X_3 are exogenous (shift) variables.]

(a) Demand: $Q = a_0 + a_1 P + a_2 X_1 + a_3 X_2 + \hat{U}$
    Supply: $Q = b_0 + b_1 P + b_2 X_3 + \hat{V}$

(b) Demand: $Q = a_0 + a_1 P + a_2 X_1 + \hat{U}$
    Supply: $Q = b_0 + b_1 P + b_2 X_3 + \hat{V}$

A: Price and quantity are simultaneously determined endogenous variables. The parameters of the supply and demand equation cannot be estimated without simultaneous equations bias if we use ordinary least-squares. With indirect least squares, however, we can obtain estimates of the parameters of the structural (behavioral) equations without simultaneous equations bias. The procedure requires computing one *reduced-form equation* for each endogenous variable and using the coefficients of the reduced-form equations to compute the parameters of the structural equations. A reduced-form equation expresses one endogenous variable as a function of the exogenous variables in the model. It will be seen that ILS can be used only when all the structural equations are exactly identified.

(a) In the first set of supply and demand equations the demand curve is exactly identified and the supply curve is over-identified. The structural equations are:

(i) Demand: $Q = a_0 + a_1 P + a_2 X_1 + a_3 X_2 + \hat{U}$
(ii) Supply: $Q = b_0 + b_1 P + b_2 X_3 + \hat{V}$

The reduced-form equations are as follows.

Quantity (Q): Converting equation (ii) into a function of P and substituting in equation (i), we get

$$Q = a_0 + \left(\frac{a_1}{b_1}\right)(Q - b_0 - b_2 X_3 - \hat{V}) + a_2 X_1 + a_3 X_2 + \hat{U},$$

$$Q\left(\frac{b_1 - a_1}{b_1}\right) = \left(\frac{b_1 a_0 - a_1 b_0}{b_1}\right) + \left(\frac{-a_1 b_2}{b_1}\right) X_3$$

$$+ a_2 X_1 + a_3 X_2 + \left(\hat{U} - \frac{a_1 \hat{V}}{b_1}\right),$$

$$Q = \left(\frac{b_1 a_0 - a_1 b_0}{b_1 - a_1}\right) + \left(\frac{-a_1 b_2}{b_1 - a_1}\right) X_3 + \left(\frac{a_2 b_1}{b_1 - a_1}\right) X_1$$

$$+ \left(\frac{a_3 b_1}{b_1 - a_1}\right) X_2 + e.$$

Then

(iii) $Q = \pi_0 + \pi_3 X_3 + \pi_1 X_1 + \pi_2 X_2 + e.$

where

$$\pi_0 = \frac{b_1 a_0 - a_1 b_0}{b_1 - a_1}, \qquad \pi_1 = \frac{a_2 b_1}{b_1 - a_1},$$

$$\pi_2 = \frac{a_3 b_1}{b_1 - a_1}, \qquad \pi_3 = \frac{-a_1 b_2}{b_1 - a_1}.$$

Equation (iii) is the reduced-form equation for Q.

Price (P): Converting equation (ii) into a function of P and substituting equation (i), we have

$$P = \left(\frac{1}{b_1}\right)(Q - b_0 - b_2 X_3 - \hat{V}),$$

$$P = \left(\frac{1}{b_1}\right)(a_0 + a_1 P + a_2 X_1 + a_3 X_2 + \hat{U} - b_0 - b_2 X_3 - \hat{V}),$$

$$P\left(\frac{b_1 - a_1}{b_1}\right) = \left(\frac{1}{b_1}\right)(a_0 - b_0 + a_2 X_1 + a_3 X_2 - b_2 X_3 + \hat{U} - \hat{V}),$$

$$P = \left(\frac{a_0 - b_0}{b_1 - a_1}\right) + \left(\frac{a_2}{b_1 - a_1}\right) X_1 + \left(\frac{a_3}{b_1 - a_1}\right) X_2$$

$$+ \left(\frac{-b_2}{b_1 - a_1}\right) X_3 + e'.$$

Then

(iv) $P = \pi_0' + \pi_1' X_1 + \pi_2' X_2 + \pi_3' X_3 + e',$

where

$$\pi_0' = \frac{a_0 - b_0}{b_1 - a_1}, \qquad \pi_1' = \frac{a_2}{b_1 - a_1}$$

$$\pi_2' = \frac{a_3}{b_1 - a_1}, \qquad \pi_3' = \frac{-b_2}{b_1 - a_1}.$$

Equation (iv) is the reduced-form equation for P.

The coefficients of the reduced-form equations (iii) and (iv) are obtained by an ordinary least-squares regression of each of the endogenous variables on the exogenous variables. Since no endogenous variable is an explanatory variable in the reduced-form equations, there is no simultaneous equations bias. Exogenous variables that are insignificant in the reduced-form equations should be deleted from the reduced-form and structural equations and the reduced-form equations are then recomputed. Let us assume that each of the reduced-form parameters is significant.

Using the regression-estimated reduced-form coefficients and the equations relating these coefficients to the structural parameters, we obtain estimates of the structural parameters.

We can obtain a_1 by

$$\frac{\pi_3}{\pi_3{}'} = \frac{-a_1 b_2/(b_1 - a_1)}{-b_2/(b_1 - a_1)} = a_1.$$

There is only one estimate of the parameter a_1. The parameter a_1 is exactly identified.

We can obtain b_1 by

$$\frac{\pi_2}{\pi_2{}'} = \frac{a_3 b_1/(b_1 - a_1)}{a_3/(b_1 - a_1)} = b_1$$

or

$$\frac{\pi_1}{\pi_1{}'} = \frac{a_2 b_1/(b_1 - a_1)}{a_2/(b_1 - a_1)} = b_1.$$

Note that there are two estimates of b_1, $\pi_1/\pi_1{}'$ and $\pi_2/\pi_2{}'$, which will, in general, differ. Each estimate of b_1 in the supply equation is due to the effect of a different exogenous (shift) explanatory variable, X_1 and X_2, in the demand equation. The parameter b_1 is over-identified. Because b_1 is over-identified, we cannot compute unique estimates of the other parameters of the structural equations. Thus ILS cannot be used to obtain unique estimates of the structural parameters in an over-identified system.

There are, however, three solutions to the inability to use ILS because of over-identification. First, the investigator may have an independent estimate or a *priori* notion of the magnitude of b_1. This value can be used as the value of b_1 to compute the other structural parameters. The quality of the estimated model is, of course, a function of the degree of reliability in estimating b_1.

Second, the two-equation model can be made exactly identified by deleting either variable X_1 or X_2 from the demand equation. If X_1 and X_2 are both statistically significant and theoretically important shift variables, deleting either variable reduces the statistical efficiency of the model and the model's usefulness to the investigator.

Third, a procedure that permits the estimation of the structural parameters from an over-identified system of equations can be used. Two-stage least squares (2SLS) is such a procedure.

(b) The system of equations in part (b) of the problem is the same as in part (a), except for the exclusion of variable X_2. Even if variable X_2 is believed on the basis of a *priori* reasoning to be an important shift variable in the demand equation, it can be excluded from the system because it is found to be statistically insignificant in the reduced-form equations or because the investigator wants to use ILS and an exogenous variable must be excluded from the demand equation.

The structural equations are:

(i) Demand: $Q = a_0 + a_1 P + a_2 X_1 + \hat{U}$

(ii) Supply: $Q = b_0 + b_1 P + b_2 X_3 + \hat{V}$

The reduced-form equations are as follows.

Quantity (Q):   (iii) $Q = \pi_0 + \pi_1 X_1 + \pi_3 X_3 + e$, where

$$\pi_0 = \frac{b_1 a_0 - a_1 b_0}{b_1 - a_1}, \qquad \pi_1 = \frac{a_2 b_1}{b_1 - a_1}, \qquad \pi_3 = \frac{-a_1 b_2}{b_1 - a_1}.$$

Price (P):   (iv) $P = \pi_0' + \pi_1' X_1 + \pi_3' X_3 + e'$, where

$$\pi_0' = \frac{a_0 - b_0}{b_1 - a_1}, \qquad \pi_1' = \frac{a_2}{b_1 - a_1}, \qquad \pi_3' = \frac{-b_2}{b_1 - a_1}.$$

The structural parameters are

$$a_1 = \frac{\pi_3}{\pi_3{}'}, \qquad b_1 = \frac{\pi_1}{\pi_1{}'},$$

$$a_2 = \pi_1{}'(b_1 - a_1) = \pi_1{}'\left(\frac{\pi_1}{\pi_1{}'} - \frac{\pi_3}{\pi_3{}'}\right),$$

$$b_2 = -\pi_3{}'(b_1 - a_1) = \pi_3{}'\left(\frac{\pi_3}{\pi_3{}'} - \frac{\pi_1}{\pi_1{}'}\right),$$

and

$$\pi_0 = \frac{b_1 a_0 - a_1 b_0}{b_1 - a_1}, \qquad \pi_0{}' = \frac{a_0 - b_0}{b_1 - a_1}.$$

These are the estimates of the structural parameters. The ILS procedure does not generate standard errors for the structural parameters.

As an example, suppose the multiple regressions for the reduced-form equations are

$$Q = 10.0 + (4.7)X_1 + (-2.5)X_3 + e,$$

$$P = 12.1 + (1.9)X_1 + (1.5)X_3 + e'.$$

The slope coefficients of the structural equations are

$$a_1 = \frac{\pi_3}{\pi_3{}'} = \frac{-2.5}{1.5} = -1.7,$$

$$b_1 = \frac{\pi_1}{\pi_1{}'} = \frac{4.7}{1.9} = 2.5,$$

$$a_2 = \pi_1{}'(b_1 - a_1) = (1.9)(2.5 + 1.7) = 8.0,$$

$$b_2 = -\pi_3{}'(b_1 - a_1) = -1.5(2.5 + 1.7) = -6.3.$$

Solving for a_0 and b_0 simultaneously, $a_0 = 30.6$ and $b_0 = -20.3$.

Problem 4. Two-Stage Least Squares

Q: Use two-stage least squares (2SLS) to compute the coefficients of a two-equation supply and demand system of equations. The structural equations in the population are:

Demand: $Q = \alpha_0 + \alpha_1 P + \alpha_2 X_1 + \alpha_3 X_2 + \alpha_4 X_3 + U,$
Supply: $P = \beta_0 + \beta_1 Q + \beta_2 X_1 + \beta_3 X_4 + V,$

where P is price; Q is quantity; X_1, X_2, X_3, X_4 are the exogenous variables which shift either the demand equation, the supply equation, or both equations; U and V are random residuals; and α and β designate the population parameters.

A: Two-stage least squares (2SLS) is a procedure for computing the coefficients of the structural equations in a simultaneous system of equations. The parameters of a structural equation can be estimated only if the equation is exactly identified or over-identified. In this problem the demand equation is exactly identified and the supply equation is over-identified (see Problem 2 of this chapter).

The slope coefficient of price in the demand equation computed from a single-equation ordinary least-squares (OLS) regression is subject to simultaneous equations bias because price and quantity determine each other.

The demand equation could be estimated if the explanatory variable price were cleansed of the effect of quantity on price. This can be done by using a *predicted* price $(\hat{P})$ rather than the observed price, if the predicted price is statistically independent of quantity (Q).

The first stage of the analysis is the computation of the ordinary least-squares regression of price (P) on *all* of the exogenous variables in the model. This equation is called the *reduced-form equation* for price. By definition, the exogenous variables are determined independently of price and quantity, so the reduced-form equation is not subject to simultaneous equations bias.

The reduced-form equation for price is written as

$$P = \pi_0 + \pi_1 X_1 + \pi_2 X_2 + \pi_3 X_3 + \pi_4 X_4 + e,$$

where the π's are the reduced-form equation coefficients and e is a random residual. The observed values of the exogenous variables are inserted into the computed reduced-form equation to obtain predicted values for price.

The predicted values of price are then used as the price data in the demand equation, and the OLS regression for the demand equation is computed. The second-stage equation is

$$Q = a_0 + a_1(\hat{P}) + a_2X_1 + a_3X_2 + a_4X_3 + \hat{U}.$$

For very large samples, a_1 is an unbiased estimate of the effect of price on quantity in the demand equation. The coefficient a_1 is a *consistent estimate* of the population coefficient α_1.

It is now clear why a two-stage least-squares analysis cannot be used to estimate the parameters of an under-identified structural equation. If the demand equation is under-identified, there is no exogenous variable in the system that is not in the demand equation. Predicted price $(\hat{P})$, however, is a perfect linear function of *all* of the exogenous variables in the system. Predicted price would then be perfectly linearly related to other variables in the demand equation. This is perfect multicollinearity, and unique estimates cannot be computed for the coefficients of the demand equation (see the introduction to Chapter 6).

Thus 2SLS can be used only if the structural equation is exactly identified or over-identified. In addition, the exogenous variables responsible for identification must be statistically significant in the reduced-form equation and sufficiently important for explaining variations in observed price so that predicted price is not "too highly" correlated with the exogenous variables in the demand equation.

A similar analysis is performed for the supply equation. If the supply equation is estimated by an ordinary least-squares regression, the slope coefficient of Q is subject to a simultaneous equations bias. There is no simultaneous equations bias, however, if quantity is cleansed of the effect of price on quantity. The reduced-form equation for quantity (first-stage regression) is

$$Q = \pi_0' + \pi_1'X_1 + \pi_2'X_2 + \pi_3'X_3 + \pi_4'X_4 + e'.$$

Predicted values for quantity $(\hat{Q})$ are obtained by inserting the observed values of the exogenous variables into the reduced-form equation. The coefficients of the supply equation are then obtained from the regression of price on predicted quantity and the exogenous variables in the supply equation:

$$P = b_0 + b_1(\hat{Q}) + b_2X_1 + b_3X_4 + \hat{V}.$$

For very large samples the coefficient b_1 is an unbiased estimate (consistent estimate) of the relation between quantity and price in the supply equation.

If all of the equations in a simultaneous system are exactly identified, the estimates of the structural coefficients obtained from the two-stage least-squares procedure are identical to those obtained from indirect least squares.

Note: The system of structural equations in this problem could have been written as:

Demand: $Q = \alpha_0{}^* + \alpha_1{}^*P + \alpha_2{}^*X_1 + \alpha_3{}^*X_2 + \alpha_4{}^*X_3 + U^*,$
Supply: $Q = \beta_0{}^* + \beta_1{}^*P + \beta_2{}^*X_1 + \beta_3{}^*X_4 + V^*.$

The reduced-form equation for price is computed from regressing price on all of the exogenous variables:

$$P = \pi_0{}^* + \pi_1{}^*X_1 + \pi_2{}^*X_2 + \pi_3{}^*X_3 + \pi_4{}^*X_4 + e^*.$$

Predicted values for price $(\hat{P})$ are obtained by using the observed values of the exogenous variables and the estimated reduced-form equation. The coefficients of the structural equations are then computed from the OLS regressions using the predicted value of the endogenous explanatory variable price $(\hat{P})$. The estimated structural equations are:

Demand: $Q = a_0{}^* + a_1{}^*\hat{P} + a_2{}^*X_1 + a_3{}^*X_2 + a_4{}^*X_3 + \hat{U}^*,$
Supply: $Q = b_0{}^* + b_1{}^*\hat{P} + b_2{}^*X_1 + b_3{}^*X_4 + \hat{V}^*.$

For very large samples the coefficients $a_1{}^*$ and $b_1{}^*$ are unbiased estimates (consistent estimates) of the effects of price on quantity in the demand equation and the supply equation, respectively.

Problem 5. Instrumental Variables

Q: Are there situations in which a two-stage regression procedure is used even though there is a single-equation model?

A: *Instrumental variables* is a statistical procedure that involves a two-stage regression analysis even though there is a single-equation model. This technique can be used when an explanatory variable is known to be measured with considerable error. It can also be used when data on an explanatory variable are not available for the same observations as the dependent variable.

1. An explanatory variable X_1 is known to be measured with considerable random error. The estimated slope coefficient b_1 computed from an ordinary least-squares regression using observed values of X_1 is a downward biased estimate of the true population parameter (β_1), where

$$Y_i = \beta_0 + \beta_1 X_{1,i} + U_i$$

(see Problem 9 of Chapter 5).

Suppose the j exogenous variables $Z_1, \ldots, Z_j$ are good predictors of the variable X_1. The OLS regression of X_1 on the instruments,

$$X_1 = c_0 + c_1 Z_1 + \cdots + c_j Z_j + \hat{V},$$

can be computed and is referred to as an *auxiliary regression*. Predicted values of X_1 are obtained by inserting observed values of the instruments into the estimated auxiliary regression equation. The predicted values $\hat{X}$ are then used in the regression equation,

$$Y = b_0' + b_1' \hat{X}_1 + \hat{U}'.$$

The OLS regression coefficient b_1' is an unbiased estimate of the population parameter β_1 when the sample is very large.

2. An investigator wishes to study the partial effect of several variables, including X_1, on a dependent variable (Y) using county data and the county as the unit of observation. Data on X_1 do not exist on a county basis, but do exist on a statewide basis. One solution is to assign to county i the value of variable X_1 for the state in which county i is located. A second solution is to use a predicted value of X_1.

Suppose variable X_1 is highly correlated with a set of exogenous variables $Z_1, \ldots, Z_j$, and data on these variables (instruments) exist at both the county and state level. The coefficients of the OLS regression,

$$X_1 = c_0 + c_1Z_1 + \cdots + c_jZ_j + \hat{V},$$

are computed using state data. The same structural relation between X_1 and the instrumental variables is assumed to exist at the county and at the state level of aggregation. Then unbiased predicted county values for X_1 can be obtained by inserting county values for the instruments into the estimated auxiliary regression. The predicted county values for X_1 are used in the OLS regression analysis of the dependent variable Y,

$$Y = b_0 + b_1(\hat{X}_1) + b_2X_2 + \cdots + b_kX_k + \hat{U}.$$

For very large samples b_1 is an unbiased estimate of the coefficient of X_1 in the population.

Note, however, that the selection of the instruments is essentially arbitrary, and different sets of instruments will result in different estimates of the structural coefficients. The reliability of the procedure depends on the extent to which the set of instruments can explain variations in X_1.

Statistical Appendix

TABLE 1   The standardized normal distribution, $Z = \dfrac{X - \mu}{\sigma}$

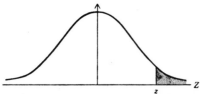

The table plots the cumulative probability $Z \geq z$. (One tail).
For an explanation and applications of the standardized normal distribution see the Introduction to Chapter 2 and Chapter 2, Problems 30 to 33.

z	.00	.01	.02	.03	.04	.05	.06	.07	.08	.09
0.0	.5000	.4960	.4920	.4880	.4840	.4801	.4761	.4721	.4681	.4641
0.1	.4602	.4562	.4522	.4483	.4443	.4404	.4364	.4325	.4286	.4247
0.2	.4207	.4168	.4129	.4090	.4052	.4013	.3974	.3936	.3897	.3859
0.3	.3821	.3783	.3745	.3707	.3669	.3632	.3594	.3557	.3520	.3483
0.4	.3446	.3409	.3372	.3336	.3300	.3264	.3228	.3192	.3156	.3121
0.5	.3085	.3050	.3015	.2981	.2946	.2912	.2877	.2843	.2810	.2776
0.6	.2743	.2709	.2676	.2643	.2611	.2578	.2546	.2514	.2483	.2451
0.7	.2420	.2389	.2358	.2327	.2296	.2266	.2236	.2206	.2177	.2148
0.8	.2119	.2090	.2061	.2033	.2005	.1977	.1949	.1922	.1894	.1867
0.9	.1841	.1814	.1788	.1762	.1736	.1711	.1685	.1660	.1635	.1611
1.0	.1587	.1562	.1539	.1515	.1492	.1469	.1446	.1423	.1401	.1379
1.1	.1357	.1335	.1314	.1292	.1271	.1251	.1230	.1210	.1190	.1170
1.2	.1151	.1131	.1112	.1093	.1075	.1056	.1038	.1020	.1003	.0985
1.3	.0968	.0951	.0934	.0918	.0901	.0885	.0869	.0853	.0838	.0823
1.4	.0808	.0793	.0778	.0764	.0749	.0735	.0721	.0708	.0694	.0681
1.5	.0668	.0655	.0643	.0630	.0618	.0606	.0594	.0582	.0571	.0559
1.6	.0548	.0537	.0526	.0516	.0505	.0495	.0485	.0475	.0465	.0455
1.7	.0446	.0436	.0427	.0418	.0409	.0401	.0392	.0384	.0375	.0367
1.8	.0359	.0351	.0344	.0336	.0329	.0322	.0314	.0307	.0301	.0294
1.9	.0287	.0281	.0274	.0268	.0262	.0256	.0250	.0244	.0239	.0233
2.0	.0228	.0222	.0217	.0212	.0207	.0202	.0197	.0192	.0188	.0183
2.1	.0179	.0174	.0170	.0166	.0162	.0158	.0154	.0150	.0146	.0143
2.2	.0139	.0136	.0132	.0129	.0125	.0122	.0119	.0116	.0113	.0110
2.3	.0107	.0104	.0102	.0099	.0096	.0094	.0091	.0089	.0087	.0084
2.4	.0082	.0080	.0078	.0075	.0073	.0071	.0069	.0068	.0066	.0064
2.5	.0062	.0060	.0059	.0057	.0055	.0054	.0052	.0051	.0049	.0048
2.6	.0047	.0045	.0044	.0043	.0041	.0040	.0039	.0038	.0037	.0036
2.7	.0035	.0034	.0033	.0032	.0031	.0030	.0029	.0028	.0027	.0026
2.8	.0026	.0025	.0024	.0023	.0023	.0022	.0021	.0021	.0020	.0019
2.9	.0019	.0018	.0018	.0017	.0016	.0016	.0015	.0015	.0014	.0014
3.0	.0013	.0013	.0013	.0012	.0012	.0011	.0011	.0011	.0010	.0010

Reprinted from Table 1, Appendix B, of Teh-wei Hu: *Econometrics, An Introductory Analysis*, 1973, University Park Press, Baltimore.

TABLE 2   Student's t Distribution

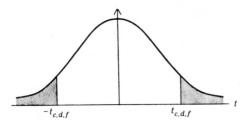

For an explanation and applications of the t-distribution see the Introduction to Chapter 3 and Chapter 3, Problems 1 and 2.

Degrees of freedom	Probability of a value greater in absolute value than the table entry (one tail)					
	0.005	0.01	0.025	0.05	0.1	0.15
1	63.657	31.821	12.706	6.314	3.078	1.963
2	9.925	6.965	4.303	2.920	1.886	1.386
3	5.841	4.541	3.182	2.353	1.638	1.250
4	4.604	3.747	2.776	2.132	1.533	1.190
5	4.032	3.365	2.571	2.015	1.476	1.156
6	3.707	3.143	2.447	1.943	1.440	1.134
7	3.499	2.998	2.365	1.895	1.415	1.119
8	3.355	2.896	2.306	1.860	1.397	1.108
9	3.250	2.821	2.262	1.833	1.383	1.100
10	3.169	2.764	2.228	1.812	1.372	1.093
11	3.106	2.718	2.201	1.796	1.363	1.088
12	3.055	2.681	2.179	1.782	1.356	1.083
13	3.012	2.650	2.160	1.771	1.350	1.079
14	2.977	2.624	2.145	1.761	1.345	1.076
15	2.947	2.602	2.131	1.753	1.341	1.074
16	2.921	2.583	2.120	1.746	1.337	1.071
17	2.898	2.567	2.110	1.740	1.333	1.069
18	2.878	2.552	2.101	1.734	1.330	1.067
19	2.861	2.539	2.093	1.729	1.328	1.066
20	2.845	2.528	2.086	1.725	1.325	1.064
21	2.831	2.518	2.080	1.721	1.323	1.063
22	2.819	2.508	2.074	1.717	1.321	1.061
23	2.807	2.500	2.069	1.714	1.319	1.060
24	2.797	2.492	2.064	1.711	1.318	1.059
25	2.787	2.485	2.060	1.708	1.316	1.058
26	2.779	2.479	2.056	1.706	1.315	1.058
27	2.771	2.473	2.052	1.703	1.314	1.057
28	2.763	2.467	2.048	1.701	1.313	1.056
29	2.756	2.462	2.045	1.699	1.311	1.055
30	2.750	2.457	2.042	1.697	1.310	1.055
∞	2.576	2.326	1.960	1.645	1.282	1.036

Reprinted from Table IV in Sir Ronald A. Fisher: *Statistical Methods for Research Workers*, 13th edition, Hafner Press. (Previously published by Oliver & Boyd, Ltd., Edinburgh, © 1963) by permission of the publishers.

TABLE 3   Critical values for the F distribution

$f(F; \eta_1, \eta_2)$

$F_{0.05}$ $F_{0.01}$ F

5% (Roman Type) and 1% (Bold Face Type) Points for the Distribution of F.
For an explanation and application of the F distribution see the Introduction to Chapter 4 and Chapter 4, Problems 3 and 4.

η_1 degrees of freedom in the numerator (each cell: roman = 5% / bold = 1%)

η_2	1	2	3	4	5	6	7	8	9	10	11	12	14	16	20	24	30	40	50	75	100	200	500	∞
1	161 / 4,052	200 / 4,999	216 / 5,403	225 / 5,625	230 / 5,764	234 / 5,859	237 / 5,928	239 / 5,981	241 / 6,022	242 / 6,056	243 / 6,082	244 / 6,106	245 / 6,142	246 / 6,169	248 / 6,208	249 / 6,234	250 / 6,258	251 / 6,286	252 / 6,302	253 / 6,323	253 / 6,334	254 / 6,352	254 / 6,361	254 / 6,366
2	18.51 / 98.49	19.00 / 99.00	19.16 / 99.17	19.25 / 99.25	19.30 / 99.30	19.33 / 99.33	19.36 / 99.34	19.37 / 99.36	19.38 / 99.38	19.39 / 99.40	19.40 / 99.41	19.41 / 99.42	19.42 / 99.43	19.43 / 99.44	19.44 / 99.45	19.45 / 99.46	19.46 / 99.47	19.47 / 99.48	19.47 / 99.48	19.48 / 99.49	19.49 / 99.49	19.49 / 99.49	19.50 / 99.50	19.50 / 99.50
3	10.13 / 34.12	9.55 / 30.82	9.28 / 29.46	9.12 / 28.71	9.01 / 28.24	8.94 / 27.91	8.88 / 27.67	8.84 / 27.49	8.81 / 27.34	8.78 / 27.23	8.76 / 27.13	8.74 / 27.05	8.71 / 26.92	8.69 / 26.83	8.66 / 26.69	8.64 / 26.60	8.62 / 26.50	8.60 / 26.41	8.58 / 26.35	8.57 / 26.27	8.56 / 26.23	8.54 / 26.18	8.54 / 26.14	8.53 / 26.12
4	7.71 / 21.20	6.94 / 18.00	6.59 / 16.69	6.39 / 15.98	6.26 / 15.52	6.16 / 15.21	6.09 / 14.98	6.04 / 14.80	6.00 / 14.66	5.96 / 14.54	5.93 / 14.45	5.91 / 14.37	5.87 / 14.24	5.84 / 14.15	5.80 / 14.02	5.77 / 13.93	5.74 / 13.83	5.71 / 13.74	5.70 / 13.69	5.68 / 13.61	5.66 / 13.57	5.65 / 13.52	5.64 / 13.48	5.63 / 13.46
5	6.61 / 16.26	5.79 / 13.27	5.41 / 12.06	5.19 / 11.39	5.05 / 10.97	4.95 / 10.67	4.88 / 10.45	4.82 / 10.27	4.78 / 10.15	4.74 / 10.05	4.70 / 9.96	4.68 / 9.89	4.64 / 9.77	4.60 / 9.68	4.56 / 9.55	4.53 / 9.47	4.50 / 9.38	4.46 / 9.29	4.44 / 9.24	4.42 / 9.17	4.40 / 9.13	4.38 / 9.07	4.37 / 9.04	4.36 / 9.02
6	5.99 / 13.74	5.14 / 10.92	4.76 / 9.78	4.53 / 9.15	4.39 / 8.75	4.28 / 8.47	4.21 / 8.26	4.15 / 8.10	4.10 / 7.98	4.06 / 7.87	4.03 / 7.79	4.00 / 7.72	3.96 / 7.60	3.92 / 7.52	3.87 / 7.39	3.84 / 7.31	3.81 / 7.23	3.77 / 7.14	3.75 / 7.09	3.72 / 7.02	3.71 / 6.99	3.69 / 6.94	3.68 / 6.90	3.67 / 6.88
7	5.59 / 12.25	4.74 / 9.55	4.35 / 8.45	4.12 / 7.85	3.97 / 7.46	3.87 / 7.19	3.79 / 7.00	3.73 / 6.84	3.68 / 6.71	3.63 / 6.62	3.60 / 6.54	3.57 / 6.47	3.52 / 6.35	3.49 / 6.27	3.44 / 6.15	3.41 / 6.07	3.38 / 5.98	3.34 / 5.90	3.32 / 5.85	3.29 / 5.78	3.28 / 5.75	3.25 / 5.70	3.24 / 5.67	3.23 / 5.65
8	5.32 / 11.26	4.46 / 8.65	4.07 / 7.59	3.84 / 7.01	3.69 / 6.63	3.58 / 6.37	3.50 / 6.19	3.44 / 6.03	3.39 / 5.91	3.34 / 5.82	3.31 / 5.74	3.28 / 5.67	3.23 / 5.56	3.20 / 5.48	3.15 / 5.36	3.12 / 5.28	3.08 / 5.20	3.05 / 5.11	3.03 / 5.06	3.00 / 5.00	2.98 / 4.96	2.96 / 4.91	2.94 / 4.88	2.93 / 4.86
9	5.12 / 10.56	4.26 / 8.02	3.86 / 6.99	3.63 / 6.42	3.48 / 6.06	3.37 / 5.80	3.29 / 5.62	3.23 / 5.47	3.18 / 5.35	3.13 / 5.26	3.10 / 5.18	3.07 / 5.11	3.02 / 5.00	2.98 / 4.92	2.93 / 4.80	2.90 / 4.73	2.86 / 4.64	2.82 / 4.56	2.80 / 4.51	2.77 / 4.45	2.76 / 4.41	2.73 / 4.36	2.72 / 4.33	2.71 / 4.31
10	4.96 / 10.04	4.10 / 7.56	3.71 / 6.55	3.48 / 5.99	3.33 / 5.64	3.22 / 5.39	3.14 / 5.21	3.07 / 5.06	3.02 / 4.95	2.97 / 4.85	2.94 / 4.78	2.91 / 4.71	2.86 / 4.60	2.82 / 4.52	2.77 / 4.41	2.74 / 4.33	2.70 / 4.25	2.67 / 4.17	2.64 / 4.12	2.61 / 4.05	2.59 / 4.01	2.56 / 3.96	2.55 / 3.93	2.54 / 3.91
11	4.84 / 9.65	3.98 / 7.20	3.59 / 6.22	3.36 / 5.67	3.20 / 5.32	3.09 / 5.07	3.01 / 4.88	2.95 / 4.74	2.90 / 4.63	2.86 / 4.54	2.82 / 4.46	2.79 / 4.40	2.74 / 4.29	2.70 / 4.21	2.65 / 4.10	2.61 / 4.02	2.57 / 3.94	2.53 / 3.86	2.50 / 3.80	2.47 / 3.74	2.45 / 3.70	2.42 / 3.66	2.41 / 3.62	2.40 / 3.60
12	4.75 / 9.33	3.88 / 6.93	3.49 / 5.95	3.26 / 5.41	3.11 / 5.06	3.00 / 4.82	2.92 / 4.65	2.85 / 4.50	2.80 / 4.39	2.76 / 4.30	2.72 / 4.22	2.69 / 4.16	2.64 / 4.05	2.60 / 3.98	2.54 / 3.86	2.50 / 3.78	2.46 / 3.70	2.42 / 3.61	2.40 / 3.56	2.36 / 3.49	2.35 / 3.46	2.32 / 3.41	2.31 / 3.38	2.30 / 3.36
13	4.67 / 9.07	3.80 / 6.70	3.41 / 5.74	3.18 / 5.20	3.02 / 4.86	2.92 / 4.62	2.84 / 4.44	2.77 / 4.30	2.72 / 4.19	2.67 / 4.10	2.63 / 4.02	2.60 / 3.96	2.55 / 3.85	2.51 / 3.78	2.46 / 3.67	2.42 / 3.59	2.38 / 3.51	2.34 / 3.42	2.32 / 3.37	2.28 / 3.30	2.26 / 3.27	2.24 / 3.21	2.22 / 3.18	2.21 / 3.16

Reprinted by permission from *Statistical Methods* by George W. Snedecor and William C. Cochran, fifth edition © 1956, by Iowa State University Press, Ames, Iowa.

TABLE 3   (*continued*)

η_1 degrees of freedom in the numerator

η_2	1	2	3	4	5	6	7	8	9	10	11	12	14	16	20	24	30	40	50	75	100	200	500	∞
14	4.60 / 8.86	3.74 / 6.51	3.34 / 5.56	3.11 / 5.03	2.96 / 4.69	2.85 / 4.46	2.77 / 4.28	2.70 / 4.14	2.65 / 4.03	2.60 / 3.94	2.56 / 3.86	2.53 / 3.80	2.48 / 3.70	2.44 / 3.62	2.39 / 3.51	2.35 / 3.43	2.31 / 3.34	2.27 / 3.26	2.24 / 3.21	2.21 / 3.14	2.19 / 3.11	2.16 / 3.06	2.14 / 3.02	2.13 / 3.00
15	4.54 / 8.68	3.68 / 6.36	3.29 / 5.42	3.06 / 4.89	2.90 / 4.56	2.79 / 4.32	2.70 / 4.14	2.64 / 4.00	2.59 / 3.89	2.55 / 3.80	2.51 / 3.73	2.48 / 3.67	2.43 / 3.56	2.39 / 3.48	2.33 / 3.36	2.29 / 3.29	2.25 / 3.20	2.21 / 3.12	2.18 / 3.07	2.15 / 3.00	2.12 / 2.97	2.10 / 2.92	2.08 / 2.89	2.07 / 2.87
16	4.49 / 8.53	3.63 / 6.23	3.24 / 5.29	3.01 / 4.77	2.85 / 4.44	2.74 / 4.20	2.66 / 4.03	2.59 / 3.89	2.54 / 3.78	2.49 / 3.69	2.45 / 3.61	2.42 / 3.55	2.37 / 3.45	2.33 / 3.37	2.28 / 3.25	2.24 / 3.18	2.20 / 3.10	2.16 / 3.01	2.13 / 2.96	2.09 / 2.89	2.07 / 2.86	2.04 / 2.80	2.02 / 2.77	2.01 / 2.75
17	4.45 / 8.40	3.59 / 6.11	3.20 / 5.18	2.96 / 4.67	2.81 / 4.34	2.70 / 4.10	2.62 / 3.93	2.55 / 3.79	2.50 / 3.68	2.45 / 3.59	2.41 / 3.52	2.38 / 3.45	2.33 / 3.35	2.29 / 3.27	2.23 / 3.16	2.19 / 3.08	2.15 / 3.00	2.11 / 2.92	2.08 / 2.86	2.04 / 2.79	2.02 / 2.76	1.99 / 2.70	1.97 / 2.67	1.96 / 2.65
18	4.41 / 8.28	3.55 / 6.01	3.16 / 5.09	2.93 / 4.58	2.77 / 4.25	2.66 / 4.01	2.58 / 3.85	2.51 / 3.71	2.46 / 3.60	2.41 / 3.54	2.37 / 3.44	2.34 / 3.37	2.29 / 3.27	2.25 / 3.19	2.19 / 3.07	2.15 / 3.00	2.11 / 2.91	2.07 / 2.83	2.04 / 2.78	2.00 / 2.71	1.98 / 2.68	1.95 / 2.62	1.93 / 2.59	1.92 / 2.57
19	4.38 / 8.18	3.52 / 5.93	3.13 / 5.01	2.90 / 4.50	2.74 / 4.17	2.63 / 3.94	2.55 / 3.77	2.48 / 3.63	2.43 / 3.52	2.38 / 3.43	2.34 / 3.36	2.31 / 3.30	2.26 / 3.19	2.21 / 3.12	2.15 / 3.00	2.11 / 2.92	2.07 / 2.84	2.02 / 2.76	2.00 / 2.70	1.96 / 2.63	1.94 / 2.60	1.91 / 2.54	1.90 / 2.51	1.88 / 2.49
20	4.35 / 8.10	3.49 / 5.85	3.10 / 4.94	2.87 / 4.43	2.71 / 4.10	2.60 / 3.87	2.52 / 3.71	2.45 / 3.56	2.40 / 3.45	2.35 / 3.37	2.31 / 3.30	2.28 / 3.23	2.23 / 3.13	2.18 / 3.05	2.12 / 2.94	2.08 / 2.86	2.04 / 2.77	1.99 / 2.69	1.96 / 2.63	1.92 / 2.56	1.90 / 2.53	1.87 / 2.47	1.85 / 2.44	1.84 / 2.42
21	4.32 / 8.02	3.47 / 5.78	3.07 / 4.87	2.84 / 4.37	2.68 / 4.04	2.57 / 3.81	2.49 / 3.65	2.42 / 3.51	2.37 / 3.40	2.32 / 3.31	2.28 / 3.24	2.25 / 3.17	2.20 / 3.07	2.15 / 2.99	2.09 / 2.88	2.05 / 2.80	2.00 / 2.72	1.96 / 2.63	1.93 / 2.58	1.89 / 2.51	1.87 / 2.47	1.84 / 2.42	1.82 / 2.38	1.81 / 2.36
22	4.30 / 7.94	3.44 / 5.72	3.05 / 4.82	2.82 / 4.31	2.66 / 3.99	2.55 / 3.76	2.47 / 3.59	2.40 / 3.45	2.35 / 3.35	2.30 / 3.26	2.26 / 3.18	2.23 / 3.12	2.18 / 3.02	2.13 / 2.94	2.07 / 2.83	2.03 / 2.75	1.98 / 2.67	1.93 / 2.58	1.91 / 2.53	1.87 / 2.46	1.84 / 2.42	1.81 / 2.37	1.80 / 2.33	1.78 / 2.31
23	4.28 / 7.88	3.42 / 5.66	3.03 / 4.76	2.80 / 4.26	2.64 / 3.94	2.53 / 3.71	2.45 / 3.54	2.38 / 3.41	2.32 / 3.30	2.28 / 3.21	2.24 / 3.14	2.20 / 3.07	2.14 / 2.97	2.10 / 2.89	2.04 / 2.78	2.00 / 2.70	1.96 / 2.62	1.91 / 2.53	1.88 / 2.48	1.84 / 2.41	1.82 / 2.37	1.79 / 2.32	1.77 / 2.28	1.76 / 2.26
24	4.26 / 7.82	3.40 / 5.61	3.01 / 4.72	2.78 / 4.22	2.62 / 3.90	2.51 / 3.67	2.43 / 3.50	2.36 / 3.36	2.30 / 3.25	2.26 / 3.17	2.22 / 3.09	2.18 / 3.03	2.13 / 2.93	2.09 / 2.85	2.02 / 2.74	1.98 / 2.66	1.94 / 2.58	1.89 / 2.49	1.86 / 2.44	1.82 / 2.36	1.80 / 2.33	1.76 / 2.27	1.74 / 2.23	1.73 / 2.21
25	4.24 / 7.77	3.38 / 5.57	2.99 / 4.68	2.76 / 4.18	2.60 / 3.86	2.49 / 3.63	2.41 / 3.46	2.34 / 3.32	2.28 / 3.21	2.24 / 3.13	2.20 / 3.05	2.16 / 2.99	2.11 / 2.89	2.06 / 2.81	2.00 / 2.70	1.96 / 2.62	1.92 / 2.54	1.87 / 2.45	1.84 / 2.40	1.80 / 2.32	1.77 / 2.29	1.74 / 2.23	1.72 / 2.19	1.71 / 2.17
26	4.22 / 7.72	3.37 / 5.53	2.98 / 4.64	2.74 / 4.14	2.59 / 3.82	2.47 / 3.59	2.39 / 3.42	2.32 / 3.29	2.27 / 3.17	2.22 / 3.09	2.18 / 3.02	2.15 / 2.96	2.10 / 2.86	2.05 / 2.77	1.99 / 2.66	1.95 / 2.58	1.90 / 2.50	1.85 / 2.41	1.82 / 2.36	1.78 / 2.28	1.76 / 2.25	1.72 / 2.19	1.70 / 2.15	1.69 / 2.13

TABLE 3 (continued)

		η_1 degrees of freedom in the numerator																						
η_2	1	2	3	4	5	6	7	8	9	10	11	12	14	16	20	24	30	40	50	75	100	200	500	∞	η_2
27	4.21 / 7.68	3.35 / 5.49	2.96 / 4.60	2.73 / 4.11	2.57 / 3.79	2.46 / 3.56	2.37 / 3.39	2.30 / 3.26	2.25 / 3.14	2.20 / 3.06	2.16 / 2.98	2.13 / 2.93	2.08 / 2.83	2.03 / 2.74	1.97 / 2.63	1.93 / 2.55	1.88 / 2.47	1.84 / 2.38	1.80 / 2.33	1.76 / 2.25	1.74 / 2.21	1.71 / 2.16	1.68 / 2.12	1.67 / 2.10	27
28	4.20 / 7.64	3.34 / 5.45	2.95 / 4.57	2.71 / 4.07	2.56 / 3.76	2.44 / 3.53	2.36 / 3.36	2.29 / 3.23	2.24 / 3.11	2.19 / 3.03	2.15 / 2.95	2.12 / 2.90	2.06 / 2.80	2.02 / 2.71	1.96 / 2.60	1.91 / 2.52	1.87 / 2.44	1.81 / 2.35	1.78 / 2.30	1.75 / 2.22	1.72 / 2.18	1.69 / 2.13	1.67 / 2.09	1.65 / 2.06	28
29	4.18 / 7.60	3.33 / 5.42	2.93 / 4.54	2.70 / 4.04	2.54 / 3.73	2.43 / 3.50	2.35 / 3.33	2.28 / 3.20	2.22 / 3.08	2.18 / 3.00	2.14 / 2.92	2.10 / 2.87	2.05 / 2.77	2.00 / 2.68	1.94 / 2.57	1.90 / 2.49	1.85 / 2.41	1.80 / 2.32	1.77 / 2.27	1.73 / 2.19	1.71 / 2.15	1.68 / 2.10	1.65 / 2.06	1.64 / 2.03	29
30	4.17 / 7.56	3.32 / 5.39	2.92 / 4.51	2.69 / 4.02	2.53 / 3.70	2.42 / 3.47	2.34 / 3.30	2.27 / 3.17	2.21 / 3.06	2.16 / 2.98	2.12 / 2.90	2.09 / 2.84	2.04 / 2.74	1.99 / 2.66	1.93 / 2.55	1.89 / 2.47	1.84 / 2.38	1.79 / 2.29	1.76 / 2.24	1.72 / 2.16	1.69 / 2.13	1.66 / 2.07	1.64 / 2.03	1.62 / 2.01	30
32	4.15 / 7.50	3.30 / 5.34	2.90 / 4.46	2.67 / 3.97	2.51 / 3.66	2.40 / 3.42	2.32 / 3.25	2.25 / 3.12	2.19 / 3.01	2.14 / 2.94	2.10 / 2.86	2.07 / 2.80	2.02 / 2.70	1.97 / 2.62	1.91 / 2.51	1.86 / 2.42	1.82 / 2.34	1.76 / 2.25	1.74 / 2.20	1.69 / 2.12	1.67 / 2.08	1.64 / 2.02	1.61 / 1.98	1.59 / 1.96	32
34	4.13 / 7.44	3.28 / 5.29	2.88 / 4.42	2.65 / 3.93	2.49 / 3.61	2.38 / 3.38	2.30 / 3.21	2.23 / 3.08	2.17 / 2.97	2.12 / 2.89	2.08 / 2.82	2.05 / 2.76	2.00 / 2.66	1.95 / 2.58	1.89 / 2.47	1.84 / 2.38	1.80 / 2.30	1.74 / 2.21	1.71 / 2.15	1.67 / 2.08	1.64 / 2.04	1.61 / 1.98	1.59 / 1.94	1.57 / 1.91	34
36	4.11 / 7.39	3.26 / 5.25	2.86 / 4.38	2.63 / 3.89	2.48 / 3.58	2.36 / 3.35	2.28 / 3.18	2.21 / 3.04	2.15 / 2.94	2.10 / 2.86	2.06 / 2.78	2.03 / 2.72	1.98 / 2.62	1.93 / 2.54	1.87 / 2.43	1.82 / 2.35	1.78 / 2.26	1.72 / 2.17	1.69 / 2.12	1.65 / 2.04	1.62 / 2.00	1.59 / 1.94	1.56 / 1.90	1.55 / 1.87	36
38	4.10 / 7.35	3.25 / 5.21	2.85 / 4.34	2.62 / 3.86	2.46 / 3.54	2.35 / 3.32	2.26 / 3.15	2.19 / 3.02	2.14 / 2.91	2.09 / 2.82	2.05 / 2.75	2.02 / 2.69	1.96 / 2.59	1.92 / 2.51	1.85 / 2.40	1.80 / 2.32	1.76 / 2.22	1.71 / 2.14	1.67 / 2.08	1.63 / 2.00	1.60 / 1.97	1.57 / 1.90	1.54 / 1.86	1.53 / 1.84	38
40	4.08 / 7.31	3.23 / 5.18	2.84 / 4.31	2.61 / 3.83	2.45 / 3.51	2.34 / 3.29	2.25 / 3.12	2.18 / 2.99	2.12 / 2.88	2.07 / 2.80	2.04 / 2.73	2.00 / 2.66	1.95 / 2.56	1.90 / 2.49	1.84 / 2.37	1.79 / 2.29	1.74 / 2.20	1.69 / 2.11	1.66 / 2.05	1.61 / 1.97	1.59 / 1.94	1.55 / 1.88	1.53 / 1.84	1.51 / 1.81	40
42	4.07 / 7.27	3.22 / 5.15	2.83 / 4.29	2.59 / 3.80	2.44 / 3.49	2.32 / 3.26	2.24 / 3.10	2.17 / 2.96	2.11 / 2.86	2.06 / 2.77	2.02 / 2.70	1.99 / 2.64	1.94 / 2.54	1.89 / 2.46	1.82 / 2.35	1.78 / 2.26	1.73 / 2.17	1.68 / 2.08	1.64 / 2.02	1.60 / 1.94	1.57 / 1.91	1.54 / 1.85	1.51 / 1.80	1.49 / 1.78	42
44	4.06 / 7.24	3.21 / 5.12	2.82 / 4.26	2.58 / 3.78	2.43 / 3.46	2.31 / 3.24	2.23 / 3.07	2.16 / 2.94	2.10 / 2.84	2.05 / 2.75	2.01 / 2.68	1.98 / 2.62	1.92 / 2.52	1.88 / 2.44	1.81 / 2.32	1.76 / 2.24	1.72 / 2.15	1.66 / 2.06	1.63 / 2.00	1.58 / 1.92	1.56 / 1.88	1.52 / 1.82	1.50 / 1.78	1.48 / 1.75	44
46	4.05 / 7.21	3.20 / 5.10	2.81 / 4.24	2.57 / 3.76	2.42 / 3.44	2.30 / 3.22	2.22 / 3.05	2.14 / 2.92	2.09 / 2.82	2.04 / 2.73	2.00 / 2.66	1.97 / 2.60	1.91 / 2.50	1.87 / 2.42	1.80 / 2.30	1.75 / 2.22	1.71 / 2.13	1.65 / 2.04	1.62 / 1.98	1.57 / 1.90	1.54 / 1.86	1.51 / 1.80	1.48 / 1.76	1.46 / 1.72	46
48	4.04 / 7.19	3.19 / 5.08	2.80 / 4.22	2.56 / 3.74	2.41 / 3.42	2.30 / 3.20	2.21 / 3.04	2.14 / 2.90	2.08 / 2.80	2.03 / 2.71	1.99 / 2.64	1.96 / 2.58	1.90 / 2.48	1.86 / 2.40	1.79 / 2.28	1.74 / 2.20	1.70 / 2.11	1.64 / 2.02	1.61 / 1.96	1.56 / 1.88	1.53 / 1.84	1.50 / 1.78	1.47 / 1.73	1.45 / 1.70	48

246 Appendix

TABLE 3 (continued)

η₁ degrees of freedom in the numerator

Each cell lists two values (upper = 5%, lower = 1% significance level).

η_2	1	2	3	4	5	6	7	8	9	10	11	12	14	16	20	24	30	40	50	75	100	200	500	∞
50	4.03/7.17	3.18/5.06	2.79/4.20	2.56/3.72	2.40/3.41	2.29/3.18	2.20/3.02	2.13/2.88	2.07/2.78	2.02/2.70	1.98/2.62	1.95/2.56	1.90/2.46	1.85/2.39	1.78/2.26	1.74/2.18	1.69/2.10	1.63/2.00	1.60/1.94	1.55/1.86	1.52/1.82	1.48/1.76	1.46/1.71	1.44/1.68
55	4.02/7.12	3.17/5.01	2.78/4.16	2.54/3.68	2.38/3.37	2.27/3.15	2.18/2.98	2.11/2.85	2.05/2.75	2.00/2.66	1.97/2.59	1.93/2.53	1.88/2.43	1.83/2.35	1.76/2.23	1.72/2.15	1.67/2.06	1.61/1.96	1.58/1.90	1.52/1.82	1.50/1.78	1.46/1.71	1.43/1.66	1.41/1.64
60	4.00/7.08	3.15/4.98	2.76/4.13	2.52/3.65	2.37/3.34	2.25/3.12	2.17/2.95	2.10/2.82	2.04/2.72	1.99/2.63	1.95/2.56	1.92/2.50	1.86/2.40	1.81/2.32	1.75/2.20	1.70/2.12	1.65/2.03	1.59/1.93	1.56/1.87	1.50/1.79	1.48/1.74	1.44/1.68	1.41/1.63	1.39/1.60
65	3.99/7.04	3.14/4.95	2.75/4.10	2.51/3.62	2.36/3.31	2.24/3.09	2.15/2.93	2.08/2.79	2.02/2.70	1.98/2.61	1.94/2.54	1.90/2.47	1.85/2.37	1.80/2.30	1.73/2.18	1.68/2.09	1.63/2.00	1.57/1.90	1.54/1.84	1.49/1.76	1.46/1.71	1.42/1.64	1.39/1.60	1.37/1.56
70	3.98/7.01	3.13/4.92	2.74/4.08	2.50/3.60	2.35/3.29	2.23/3.07	2.14/2.91	2.07/2.77	2.01/2.67	1.97/2.59	1.93/2.51	1.89/2.45	1.84/2.35	1.79/2.28	1.72/2.15	1.67/2.07	1.62/1.98	1.56/1.88	1.53/1.82	1.47/1.74	1.45/1.69	1.40/1.62	1.37/1.56	1.35/1.53
80	3.96/6.96	3.11/4.88	2.72/4.04	2.48/3.56	2.33/3.25	2.21/3.04	2.12/2.87	2.05/2.74	1.99/2.64	1.95/2.55	1.91/2.48	1.88/2.41	1.82/2.32	1.77/2.24	1.70/2.11	1.65/2.03	1.60/1.94	1.54/1.84	1.51/1.78	1.45/1.70	1.42/1.65	1.38/1.57	1.35/1.52	1.32/1.49
100	3.94/6.90	3.09/4.82	2.70/3.98	2.46/3.51	2.30/3.20	2.19/2.99	2.10/2.82	2.03/2.69	1.97/2.59	1.92/2.51	1.88/2.43	1.85/2.36	1.79/2.26	1.75/2.19	1.68/2.06	1.63/1.98	1.57/1.89	1.51/1.79	1.48/1.73	1.42/1.64	1.39/1.59	1.34/1.51	1.30/1.46	1.28/1.43
125	3.92/6.84	3.07/4.78	2.68/3.94	2.44/3.47	2.29/3.17	2.17/2.95	2.08/2.79	2.01/2.65	1.95/2.56	1.90/2.47	1.86/2.40	1.83/2.33	1.77/2.23	1.72/2.15	1.65/2.03	1.60/1.94	1.55/1.85	1.49/1.75	1.45/1.68	1.39/1.59	1.36/1.54	1.31/1.46	1.27/1.40	1.25/1.37
150	3.91/6.81	3.06/4.75	2.67/3.91	2.43/3.44	2.27/3.14	2.16/2.92	2.07/2.76	2.00/2.62	1.94/2.53	1.89/2.44	1.85/2.37	1.82/2.30	1.76/2.20	1.71/2.12	1.64/2.00	1.59/1.91	1.54/1.83	1.47/1.72	1.44/1.66	1.37/1.56	1.34/1.51	1.29/1.43	1.25/1.37	1.22/1.33
200	3.89/6.76	3.04/4.71	2.65/3.88	2.41/3.41	2.26/3.11	2.14/2.90	2.05/2.73	1.98/2.60	1.92/2.50	1.87/2.41	1.83/2.34	1.80/2.28	1.74/2.17	1.69/2.09	1.62/1.97	1.57/1.88	1.52/1.79	1.45/1.69	1.42/1.62	1.35/1.53	1.32/1.48	1.26/1.39	1.22/1.33	1.19/1.28
400	3.86/6.70	3.02/4.66	2.62/3.83	2.39/3.36	2.23/3.06	2.12/2.85	2.03/2.69	1.96/2.55	1.90/2.46	1.85/2.37	1.81/2.29	1.78/2.23	1.72/2.12	1.67/2.04	1.60/1.92	1.54/1.84	1.49/1.74	1.42/1.64	1.38/1.57	1.32/1.47	1.28/1.42	1.22/1.32	1.16/1.24	1.13/1.19
1000	3.85/6.66	3.00/4.62	2.61/3.80	2.38/3.34	2.22/3.04	2.10/2.82	2.02/2.66	1.95/2.53	1.89/2.43	1.84/2.34	1.80/2.26	1.76/2.20	1.70/2.09	1.65/2.01	1.58/1.89	1.53/1.81	1.47/1.71	1.41/1.61	1.36/1.54	1.30/1.44	1.26/1.38	1.19/1.28	1.13/1.19	1.08/1.11
∞	3.84/6.64	2.99/4.60	2.60/3.78	2.37/3.32	2.21/3.02	2.09/2.80	2.01/2.64	1.94/2.51	1.88/2.41	1.83/2.32	1.79/2.24	1.75/2.18	1.69/2.07	1.64/1.99	1.57/1.87	1.52/1.79	1.46/1.69	1.40/1.59	1.35/1.52	1.28/1.41	1.24/1.36	1.17/1.25	1.11/1.15	1.00/1.00

TABLE 4   Critical values for the Durbin-Watson test; 5% significance
points of d_l and d_u in two-tailed tests.

For an explanation and application see Chapter 6, Problem 11.

d.f.	k = 1		k = 2		k = 3		k = 4		k = 5	
	d_l	d_u	d_l	d_u	d_l	d_u	d_l	d_u	d_l	d_u
15	0.95	1.23	0.83	1.40	0.71	1.61	0.59	1.84	0.48	2.09
16	0.98	1.24	0.86	1.40	0.75	1.59	0.64	1.80	0.53	2.03
17	1.01	1.25	0.90	1.40	0.79	1.58	0.68	1.77	0.57	1.98
18	1.03	1.26	0.93	1.40	0.82	1.56	0.72	1.74	0.62	1.93
19	1.06	1.28	0.96	1.41	0.86	1.55	0.76	1.72	0.66	1.90
20	1.08	1.28	0.99	1.41	0.89	1.55	0.79	1.70	0.70	1.87
21	1.10	1.30	1.01	1.41	0.92	1.54	0.83	1.69	0.73	1.84
22	1.12	1.31	1.04	1.42	0.95	1.54	0.86	1.68	0.77	1.82
23	1.14	1.32	1.06	1.42	0.97	1.54	0.89	1.67	0.80	1.80
24	1.16	1.33	1.08	1.43	1.00	1.54	0.91	1.66	0.83	1.79
25	1.18	1.34	1.10	1.43	1.02	1.54	0.94	1.65	0.86	1.77
26	1.19	1.35	1.12	1.44	1.04	1.54	0.96	1.65	0.88	1.76
27	1.21	1.36	1.13	1.44	1.06	1.54	0.99	1.64	0.91	1.75
28	1.22	1.37	1.15	1.45	1.08	1.54	1.01	1.64	0.93	1.74
29	1.24	1.38	1.17	1.45	1.10	1.54	1.03	1.63	0.96	1.73
30	1.25	1.38	1.18	1.46	1.12	1.54	1.05	1.63	0.98	1.73
31	1.26	1.39	1.20	1.47	1.13	1.55	1.07	1.63	1.00	1.72
32	1.27	1.40	1.21	1.47	1.15	1.55	1.08	1.63	1.02	1.71
33	1.28	1.41	1.22	1.48	1.16	1.55	1.10	1.63	1.04	1.71
34	1.29	1.41	1.24	1.48	1.17	1.55	1.12	1.63	1.06	1.70
35	1.30	1.42	1.25	1.48	1.19	1.55	1.13	1.63	1.07	1.70
36	1.31	1.43	1.26	1.49	1.20	1.56	1.15	1.63	1.09	1.70
37	1.32	1.43	1.27	1.49	1.21	1.56	1.16	1.62	1.10	1.70
38	1.33	1.44	1.28	1.50	1.23	1.56	1.17	1.62	1.12	1.70
39	1.34	1.44	1.29	1.50	1.24	1.56	1.19	1.63	1.13	1.69
40	1.35	1.45	1.30	1.51	1.25	1.57	1.20	1.63	1.15	1.69
45	1.39	1.48	1.34	1.53	1.30	1.58	1.25	1.63	1.21	1.69
50	1.42	1.50	1.38	1.54	1.34	1.59	1.30	1.64	1.26	1.69
55	1.45	1.52	1.41	1.56	1.37	1.60	1.33	1.64	1.30	1.69
60	1.47	1.54	1.44	1.57	1.40	1.61	1.37	1.65	1.33	1.69
65	1.49	1.55	1.46	1.59	1.43	1.62	1.40	1.66	1.36	1.69
70	1.51	1.57	1.48	1.60	1.45	1.63	1.42	1.66	1.39	1.70
75	1.53	1.58	1.50	1.61	1.47	1.64	1.45	1.67	1.42	1.70
80	1.54	1.59	1.52	1.62	1.49	1.65	1.47	1.67	1.44	1.70
85	1.56	1.60	1.53	1.63	1.51	1.65	1.49	1.68	1.46	1.71
90	1.57	1.61	1.55	1.64	1.53	1.66	1.50	1.69	1.48	1.71
95	1.58	1.62	1.56	1.65	1.54	1.67	1.52	1.69	1.50	1.71
100	1.59	1.63	1.57	1.65	1.55	1.67	1.53	1.70	1.51	1.72

Modified from table in J. Durbin and G. S. Watson "Testing for Serial Correlation in Least Squares Regression," *Biometrika*, Vol. 38 (1951). By permission of the authors and The Biometrika Trustees.
d.f. = degrees of freedom; k = Number of independent variables.

TABLE 5   Chi-square Distribution—Percentile Values

For an explanation and application see the Introduction to Chapter 4 and Chapter 4, Problem 1.

d.f.	$\chi^2_{.005}$	$\chi^2_{.01}$	$\chi^2_{.02}$	$\chi^2_{.025}$	$\chi^2_{.05}$	$\chi^2_{.10}$	$\chi^2_{.25}$	$\chi^2_{.50}$	$\chi^2_{.75}$	$\chi^2_{.90}$	$\chi^2_{.95}$	$\chi^2_{.975}$	$\chi^2_{.98}$	$\chi^2_{.99}$	$\chi^2_{.995}$	$\chi^2_{.999}$	d.f.
1	—	—	—	—	—	.02	.10	.46	1.3	2.7	3.8	5.0	5.4	6.6	7.9	10.8	1
2	.01	.02	.04	.05	.10	.21	.58	1.4	2.8	4.6	6.0	7.4	7.8	9.2	10.6	13.8	2
3	.07	.11	.18	.22	.35	.58	1.21	2.4	4.1	6.3	7.8	9.4	9.8	11.3	12.8	16.3	3
4	.21	.30	.43	.48	.71	1.1	1.92	3.4	5.4	7.8	9.5	11.1	11.7	13.3	14.9	18.5	4
5	.41	.55	.75	.83	1.1	1.6	2.7	4.4	6.6	9.2	11.1	12.8	13.4	15.1	16.7	20.5	5
6	.68	.87	1.13	1.2	1.6	2.2	3.5	5.4	7.8	10.6	12.6	14.4	15.0	16.8	18.5	22.5	6
7	.99	1.24	1.56	1.7	2.2	2.8	4.3	6.4	9.0	12.0	14.1	16.0	16.6	18.5	20.3	24.3	7
8	1.3	1.65	2.03	2.2	2.7	3.5	5.1	7.3	10.2	13.4	15.5	17.5	18.2	20.1	22.0	26.1	8
9	1.7	2.09	2.53	2.7	3.3	4.2	5.9	8.3	11.4	14.7	16.9	19.0	19.7	21.7	23.6	27.9	9
10	2.2	2.55	3.06	3.2	3.9	4.9	6.7	9.3	12.5	16.0	18.3	20.5	21.2	23.2	25.2	29.6	10
11	2.6	3.05	3.61	3.8	4.6	5.6	7.6	10.3	13.7	17.3	19.7	21.9	22.6	24.7	26.8	31.3	11
12	3.1	3.57	4.18	4.4	5.2	6.3	8.4	11.3	14.8	18.5	21.0	23.3	24.1	26.2	28.3	32.9	12
13	3.6	4.11	4.76	5.0	5.9	7.0	9.3	12.3	16.0	19.8	22.4	24.7	25.5	27.7	29.8	34.5	13
14	4.1	4.66	5.37	5.6	6.6	7.8	10.2	13.3	17.1	21.1	23.7	26.1	26.9	29.1	31.3	36.1	14
15	4.6	5.23	5.98	6.3	7.3	8.5	11.0	14.3	18.2	22.3	25.0	27.5	28.3	30.6	32.8	37.7	15
16	5.1	5.81	6.61	6.9	8.0	9.3	11.9	15.3	19.4	23.5	26.3	28.8	29.6	32.0	34.3	39.3	16
17	5.7	6.41	7.26	7.6	8.7	10.1	12.8	16.3	20.5	24.8	27.6	30.2	31.0	33.4	35.7	40.8	17
18	6.3	7.02	7.91	8.2	9.4	10.9	13.7	17.3	21.6	26.0	28.9	31.5	32.3	34.8	37.2	42.3	18
19	6.9	7.63	8.57	8.9	10.1	11.7	14.6	18.3	22.7	27.2	30.1	32.9	33.7	36.2	38.6	43.8	19
20	7.4	8.26	9.24	9.6	10.9	12.4	15.5	19.3	23.8	28.4	31.4	34.2	35.0	37.6	40.0	45.3	20
21	8.0	8.9	9.9	10.3	11.6	13.2	16.3	20.3	24.9	29.6	32.7	35.5	36.3	38.9	41.4	46.8	21
22	8.6	9.5	10.6	11.0	12.3	14.0	17.2	21.3	26.0	30.8	33.9	36.8	37.7	40.3	42.8	48.3	22
23	9.3	10.2	11.3	11.7	13.1	14.8	18.1	22.3	27.1	32.0	35.2	38.1	39.0	41.6	44.2	49.7	23
24	9.9	10.9	12.0	12.4	13.8	15.7	19.0	23.3	28.2	33.2	36.4	39.4	40.3	43.0	45.6	51.2	24
25	10.5	11.5	12.7	13.1	14.6	16.5	19.9	24.3	29.3	34.4	37.7	40.6	41.6	44.3	46.9	52.6	25
26	11.2	12.2	13.4	13.8	15.4	17.3	20.8	25.3	30.4	35.6	38.9	41.9	42.9	45.6	48.3	54.0	26
27	11.8	12.9	14.1	14.6	16.2	18.1	21.7	26.3	31.5	36.7	40.1	43.2	44.1	47.0	49.6	55.5	27
28	12.5	13.6	14.8	15.3	16.9	18.9	22.7	27.3	32.6	37.9	41.3	44.5	45.4	48.3	51.0	56.9	28
29	13.1	14.3	15.6	16.0	17.7	19.8	23.6	28.3	33.7	39.1	42.6	45.7	46.7	49.6	52.4	58.3	29
30	13.8	15.0	16.3	16.8	18.5	20.6	24.5	29.3	34.8	40.3	43.8	47.0	48.0	50.9	53.3	59.7	30
40	20.7	22.2	23.8	24.4	26.5	29.1	33.7	39.3	45.6	51.8	55.8	59.3	60.4	63.7	66.8	73.5	40
60	35.5	37.5	39.7	40.5	43.2	46.5	52.3	59.3	67.0	74.4	79.1	83.3	84.6	88.4	92.0	99.7	60
100	67.3	70.0	73.1	74.2	77.9	82.4	90.1	99.3	109.1	118.5	124.3	129.6	131.1	135.8	140.2	149.5	100

Abridged from *Biometrika Tables for Statistics*, Vol. 1, 3rd edition 1966. By permission of The Biometrika Trustees. Columns $\chi^2_{.02}$, $\chi^2_{.98}$, and $\chi^2_{.999}$ are taken abridged from Table 245 of Fisher and Yates: *Statistical Tables for Biological, Agricultural and Medical Research*, published by Longman Group, Ltd., London. (previously published by Oliver & Boyd, Edinburgh), and by permission of the authors and publishers.

d.f. = degrees of freedom.

TABLE 6 F_{max} Distribution—95th and 99th Percentile Values of $F_{max} = s^2_{max}/s^2_{min}$ in a Set of k Mean Squares each Based on $d.f.$ Degrees of Freedom

For an explanation and application see Chapter 4, Problem 5.

d.f. \ k	2	3	4	5	6
2	39.0	87.5	142.	202.	266.
	199.	448.	729.	1036.	1362.
3	15.4	27.8	39.2	50.7	62.0
	47.5	85.	120.	151.	184.
4	9.60	15.5	20.6	25.2	29.5
	23.2	37.	49.	59.	69.
5	7.15	10.8	13.7	16.3	18.7
	14.9	22.	28.	33.	38.
6	5.82	8.38	10.4	12.1	13.7
	11.1	15.5	19.1	22.	25.
7	4.99	6.94	8.44	9.70	10.8
	8.89	12.1	14.5	16.5	18.4
8	4.43	6.00	7.18	8.12	9.03
	7.50	9.9	11.7	13.2	14.5
9	4.03	5.34	6.31	7.11	7.80
	6.54	8.5	9.9	11.1	12.1
10	3.72	4.85	5.67	6.34	6.92
	5.85	7.4	8.6	9.6	10.4
12	3.28	4.16	4.79	5.30	5.72
	4.91	6.1	6.9	7.6	8.2
15	2.86	3.54	4.01	4.37	4.68
	4.07	4.9	5.5	6.0	6.4
20	2.46	2.95	3.29	3.54	3.76
	3.32	3.8	4.3	4.6	4.9
30	2.07	2.40	2.61	2.78	2.91
	2.63	3.0	3.3	3.4	3.6
60	1.67	1.85	1.96	2.04	2.11
	1.96	2.2	2.3	2.4	2.4
∞	1.00	1.00	1.00	1.00	1.00
	1.00	1.00	1.00	1.00	1.00

Reprinted from the table by H. A. David in *Biometrika Tables for Statistics*, Vol. 1, 3rd edition, 1966. By permission of The Biometrika Trustees. Values in the column $k = 2$ and in the rows $d.f. = 2$ and ∞ are exact. Elsewhere the third digit may be in error by several units. Upper value in each cell is the 95th percentile, lower value is the 99th. * indicates that third digit is uncertain.
k = Number of Variances; $d.f.$ = Number of Degrees of Freedom.

TABLE 6   (*continued*)

7	8	9	10	11	12	d.f.
333.	403.	475.	550.	626.	704.	2
1705.	2063.	2432.	2813.	3204.	3605.	
72.9	83.5	93.9	104.	114.	124.	3
216.*	249.*	281.*	310.*	337.*	361.*	
33.6	37.5	41.1	44.6	48.0	51.4	4
79.	89.	97.	106.	113.	120.	
20.8	22.9	24.7	26.5	28.2	29.9	5
42.	46.	50.	54.	57.	60.	
15.0	16.3	17.5	18.6	19.7	20.7	6
27.	30.	32.	34.	36.	37.	
11.8	12.7	13.5	14.3	15.1	15.8	7
20.	22.	23.	24.	26.	27.	
9.78	10.5	11.1	11.7	12.2	12.7	8
15.8	16.9	17.9	18.9	19.8	21.	
8.41	8.95	9.45	9.91	10.3	10.7	9
13.1	13.9	14.7	15.3	16.0	16.6	
7.42	7.87	8.28	8.66	9.01	9.34	10
11.1	11.8	12.4	12.9	13.4	13.9	
6.09	6.42	6.72	7.00	7.25	7.48	12
8.7	9.1	9.5	9.9	10.2	10.6	
4.95	5.19	5.40	5.59	5.77	5.93	15
6.7	7.1	7.3	7.5	7.8	8.0	
3.94	4.10	4.24	4.37	4.49	4.59	20
5.1	5.3	5.5	5.6	5.8	5.9	
3.02	3.12	3.21	3.29	3.36	3.39	30
3.7	3.8	3.9	4.0	4.1	4.2	
2.17	2.22	2.26	2.30	2.33	2.36	60
2.5	2.5	2.6	2.6	2.7	2.7	
1.00	1.00	1.00	1.00	1.00	1.00	∞
1.00	1.00	1.00	1.00	1.00	1.00	

TABLE 7   The Poisson Approximation

For an explanation and application see the Introduction to Chapter 3 and Chapter 3, Problem 8 or Chapter 2, Problem 34.

If the population mean (NP) of a Poisson Distribution is $\lambda = 4.0$, the probability of obtaining three or fewer successes is: $P(S \leq 3) = P(S = 0) + P(S = 1) + P(S = 2) + P(S = 3) = 0.0183 + 0.0733 + 0.1465 + 0.1954 = 0.4335$. (Chap. 3, Prob. 8).

S	$NP = \lambda$ .1	.2	.3	.4	.5	.6	.7	.8	.9	1.0
0	.9048	.8187	.7408	.6703	.6065	.5488	.4966	.4493	.4066	.3679
1	.0905	.1637	.2222	.2681	.3033	.3293	.3476	.3595	.3659	.3679
2	.0045	.0164	.0333	.0536	.0758	.0988	.1217	.1438	.1647	.1839
3	.0002	.0011	.0033	.0072	.0126	.0198	.0284	.0383	.0494	.0613
4		.0001	.0003	.0007	.0016	.0030	.0050	.0077	.0111	.0153
5				.0001	.0002	.0004	.0007	.0012	.0020	.0031
6							.0001	.0002	.0003	.0005
7										.0001

S	$NP = \lambda$ 1	2	3	4	5	6	7	8	9	10
0	.3679	.1353	.0498	.0183	.0067	.0025	.0009	.0003	.0001	.0000
1	.3679	.2707	.1494	.0733	.0337	.0149	.0064	.0027	.0011	.0005
2	.1839	.2707	.2240	.1465	.0842	.0446	.0223	.0107	.0050	.0023
3	.0613	.1804	.2240	.1954	.1404	.0892	.0521	.0286	.0150	.0076
4	.0153	.0902	.1680	.1954	.1755	.1339	.0912	.0572	.0337	.0189
5	.0031	.0361	.1008	.1563	.1755	.1606	.1277	.0916	.0607	.0378
6	.0005	.0120	.0504	.1042	.1462	.1606	.1490	.1221	.0911	.0631
7	.0001	.0034	.0216	.0595	.1044	.1377	.1490	.1396	.1171	.0901
8		.0009	.0081	.0298	.0653	.1033	.1304	.1396	.1318	.1126
9		.0002	.0027	.0132	.0363	.0688	.1014	.1241	.1318	.1251
10			.0008	.0053	.0181	.0413	.0710	.0993	.1186	.1251
11			.0002	.0019	.0082	.0225	.0452	.0722	.0970	.1137
12			.0001	.0006	.0034	.0113	.0264	.0481	.0728	.0948
13				.0002	.0013	.0052	.0142	.0296	.0504	.0729
14				.0001	.0005	.0022	.0071	.0169	.0324	.0521
15					.0002	.0009	.0033	.0090	.0194	.0347
16						.0003	.0014	.0045	.0109	.0217
17						.0001	.0006	.0021	.0058	.0128
18							.0002	.0009	.0029	.0071
19							.0001	.0004	.0014	.0037
20								.0002	.0006	.0019
21								.0001	.0003	.0009
22									.0001	.0004
23										.0002
24										.0001

Modified from Table F, page 417 in J. L. Hodges, Jr. and E. L. Lehmann: *Basic Concepts of Probability and Statistics*, 2nd edition, Holden-Day, Inc. Publishers. By permission of the authors and publishers.

TABLE 8   Binomial Distribution (N,P)

P = Population Proportion
N = Number of Observations
S = Number of Successes

For an explanation and applications, see Introduction to Chapter 2 and Chapter 2, Problems 27 and 33.

N	S	P = .05	P = .1	P = .2	P = .3	P = .4		N	S	P = .05	P = .1	P = .2	P = .3	P = .4
2	0	.9025	.8100	.6400	.4900	.3600		11	0	.5688	.3138	.0859	.0198	.0036
	1	.0950	.1800	.3200	.4200	.4800			1	.3293	.3835	.2362	.0932	.0266
	2	.0025	.0100	.0400	.0900	.1600			2	.0867	.2131	.2953	.1998	.0887
3	0	.8574	.7290	.5120	.3430	.2160			3	.0137	.0710	.2215	.2568	.1774
	1	.1354	.2430	.3840	.4410	.4320			4	.0014	.0158	.1107	.2201	.2365
	2	.0071	.0270	.0960	.1890	.2880			5	.0001	.0025	.0388	.1321	.2207
	3	.0001	.0010	.0080	.0270	.0640			6		.0003	.0097	.0566	.1471
4	0	.8145	.6561	.4096	.2401	.1296			7			.0017	.0173	.0701
	1	.1715	.2916	.4096	.4116	.3456			8			.0002	.0037	.0234
	2	.0135	.0486	.1536	.2646	.3456			9				.0005	.0052
	3	.0005	.0036	.0256	.0756	.1536			10					.0007
	4		.0001	.0016	.0081	.0256		12	0	.5404	.2824	.0687	.0138	.0022
5	0	.7738	.5905	.3277	.1681	.0778			1	.3413	.3766	.2062	.0712	.0174
	1	.2036	.3280	.4096	.3602	.2592			2	.0988	.2301	.2835	.1678	.0639
	2	.0214	.0729	.2048	.3087	.3456			3	.0173	.0852	.2362	.2397	.1419
	3	.0011	.0081	.0512	.1323	.2304			4	.0021	.0213	.1329	.2311	.2128
	4		.0005	.0064	.0284	.0768			5	.0002	.0038	.0532	.1585	.2270
	5			.0003	.0024	.0102			6		.0005	.0155	.0792	.1766
6	0	.7351	.5314	.2621	.1176	.0467			7			.0033	.0291	.1009
	1	.2321	.3543	.3932	.3025	.1866			8			.0005	.0078	.0420
	2	.0305	.0984	.2458	.3241	.3110			9			.0001	.0015	.0125
	3	.0021	.0146	.0819	.1852	.2765			10				.0002	.0025
	4	.0001	.0012	.0154	.0595	.1382			11					.0003
	5		.0001	.0015	.0102	.0369		13	0	.5133	.2542	.0550	.0097	.0013
	6			.0001	.0007	.0041			1	.3512	.3672	.1787	.0540	.0113
7	0	.6983	.4783	.2097	.0824	.0280			2	.1109	.2448	.2680	.1388	.0453
	1	.2573	.3720	.3670	.2471	.1306			3	.0214	.0997	.2457	.2181	.1107
	2	.0406	.1240	.2753	.3176	.2613			4	.0028	.0277	.1535	.2337	.1845
	3	.0036	.0230	.1147	.2269	.2903			5	.0003	.0055	.0691	.1803	.2214
	4	.0002	.0026	.0287	.0972	.1935			6		.0008	.0230	.1030	.1968
	5		.0002	.0043	.0250	.0774			7		.0001	.0058	.0442	.1312
	6			.0004	.0036	.0172			8			.0011	.0142	.0656
	7				.0002	.0016			9			.0001	.0034	.0243
8	0	.6634	.4305	.1678	.0576	.0168			10				.0006	.0065
	1	.2793	.3826	.3355	.1977	.0896			11				.0001	.0012
	2	.0515	.1488	.2936	.2965	.2090			12					.0001
	3	.0054	.0331	.1468	.2541	.2787		14	0	.4877	.2288	.0440	.0068	.0008
	4	.0004	.0046	.0459	.1361	.2322			1	.3593	.3559	.1539	.0407	.0073
	5		.0004	.0092	.0467	.1239			2	.1229	.2570	.2501	.1134	.0317
	6			.0011	.0100	.0413			3	.0259	.1142	.2501	.1943	.0845
	7			.0001	.0012	.0079			4	.0037	.0349	.1720	.2290	.1549
	8				.0001	.0007			5	.0004	.0078	.0860	.1963	.2066
9	0	.6302	.3874	.1342	.0404	.0101			6		.0013	.0322	.1262	.2066
	1	.2985	.3874	.3020	.1556	.0605			7		.0002	.0092	.0618	.1574
	2	.0629	.1722	.3020	.2668	.1612			8			.0020	.0232	.0918
	3	.0077	.0446	.1762	.2668	.2508			9			.0003	.0066	.0408
	4	.0006	.0074	.0661	.1715	.2508			10				.0014	.0136
	5		.0008	.0165	.0735	.1672			11				.0002	.0033
	6		.0001	.0028	.0210	.0743			12					.0005
	7			.0003	.0039	.0212			13					.0001
	8				.0004	.0035		15	0	.4633	.2059	.0352	.0047	.0005
	9					.0003			1	.3658	.3432	.1319	.0305	.0047
10	0	.5987	.3487	.1074	.0282	.0060			2	.1348	.2669	.2309	.0916	.0219
	1	.3151	.3874	.2684	.1211	.0403			3	.0307	.1285	.2501	.1700	.0634
	2	.0746	.1937	.3020	.2335	.1209			4	.0049	.0428	.1876	.2186	.1268
	3	.0105	.0574	.2013	.2668	.2150			5	.0006	.0105	.1032	.2061	.1859
	4	.0010	.0112	.0881	.2001	.2508			6		.0019	.0430	.1472	.2066
	5	.0001	.0015	.0264	.1029	.2007			7		.0003	.0138	.0811	.1771
	6		.0001	.0055	.0368	.1115			8			.0035	.0348	.1181
	7			.0008	.0090	.0425			9			.0007	.0116	.0612
	8			.0001	.0014	.0106			10			.0001	.0030	.0245
	9				.0001	.0016			11				.0006	.0074
	10					.0001			12				.0001	.0016
									13					.0003

TABLE 8   (continued) Binomial Distribution ($N,P = 0.5$)

N	S	P = .5	N	S	P = .5	N	S	P = .5	N	S	P = .5	N	S	P = .5
2	0	.2500	13	0	.0001	18	0	.0000	23	2	.0000	27	3	.0000
	1	.5000		1	.0016		1	.0001		3	.0002		4	.0001
3	0	.1250		2	.0095		2	.0006		4	.0011		5	.0006
	1	.3750		3	.0349		3	.0031		5	.0040		6	.0022
4	0	.0625		4	.0873		4	.0117		6	.0120		7	.0066
	1	.2500		5	.1571		5	.0327		7	.0292		8	.0165
	2	.3750		6	.2095		6	.0708		8	.0584		9	.0349
5	0	.0312	14	0	.0001		7	.1214		9	.0974		10	.0629
	1	.1562		1	.0009		8	.1669		10	.1364		11	.0971
	2	.3125		2	.0056		9	.1855		11	.1612		12	.1295
6	0	.0156		3	.0222	19	1	.0000	24	2	.0000		13	.1494
	1	.0938		4	.0611		2	.0003		3	.0001	28	3	.0000
	2	.2344		5	.1222		3	.0018		4	.0006		4	.0001
	3	.3125		6	.1833		4	.0074		5	.0025		5	.0004
7	0	.0078		7	.2095		5	.0222		6	.0080		6	.0014
	1	.0547	15	0	.0000		6	.0518		7	.0206		7	.0044
	2	.1641		1	.0005		7	.0961		8	.0438		8	.0116
	3	.2734		2	.0032		8	.1442		9	.0779		9	.0257
8	0	.0039		3	.0139		9	.1762		10	.1169		10	.0489
	1	.0312		4	.0417	20	1	.0000		11	.1488		11	.0800
	2	.1094		5	.0916		2	.0002		12	.1612		12	.1133
	3	.2188		6	.1527		3	.0011	25	2	.0000		13	.1395
	4	.2734		7	.1964		4	.0046		3	.0001		14	.1494
9	0	.0020	16	0	.0000		5	.0148		4	.0004	29	4	.0000
	1	.0176		1	.0002		6	.0370		5	.0016		5	.0002
	2	.0703		2	.0018		7	.0739		6	.0053		6	.0009
	3	.1641		3	.0085		8	.1201		7	.0143		7	.0029
	4	.2461		4	.0278		9	.1602		8	.0322		8	.0080
10	0	.0010		5	.0667		10	.1762		9	.0609		9	.0187
	1	.0098		6	.1222	21	1	.0000		10	.0974		10	.0373
	2	.0439		7	.1746		2	.0001		11	.1328		11	.0644
	3	.1172		8	.1964		3	.0006		12	.1550		12	.0967
	4	.2051	17	0	.0000		4	.0029	26	3	.0000		13	.1264
	5	.2461		1	.0001		5	.0097		4	.0002		14	.1445
11	0	.0005		2	.0010		6	.0259		5	.0010	30	4	.0000
	1	.0054		3	.0052		7	.0554		6	.0034		5	.0001
	2	.0269		4	.0182		8	.0970		7	.0098		6	.0006
	3	.0806		5	.0472		9	.1402		8	.0233		7	.0019
	4	.1611		6	.0944		10	.1682		9	.0466		8	.0055
	5	.2256		7	.1484	22	1	.0000		10	.0792		9	.0133
12	0	.0002		8	.1855		2	.0001		11	.1151		10	.0280
	1	.0029					3	.0001		12	.1439		11	.0509
	2	.0161					4	.0017		13	.1550		12	.0806
	3	.0537					5	.0063					13	.1115
	4	.1208					6	.0178					14	.1354
	5	.1934					7	.0407					15	.1445
	6	.2256					8	.0762						
							9	.1186						
							10	.1542						
							11	.1682						

Modified from Table B, page 413 in J. L. Hodges, Jr. and E. L. Lehmann: *Basic Concepts of Probability and Statistics*, 2nd edition, Holden-Day, Inc., Publishers. By permission of the authors and publishers.

TABLE 9 Squares, Square Roots, and Reciprocals

n	n^2	$\sqrt{n}$	$\sqrt{10n}$	$1/n$	n	n^2	$\sqrt{n}$	$\sqrt{10n}$	$1/n$
1	1	1.000	3.162	1.00000	51	2601	7.141	22.583	.01961
2	4	1.414	4.472	.50000	52	2704	7.211	22.804	.01923
3	9	1.732	5.477	.33333	53	2809	7.280	23.022	.01887
4	16	2.000	6.325	.25000	54	2916	7.348	23.238	.01852
5	25	2.236	7.071	.20000	55	3025	7.416	23.452	.01818
6	36	2.449	7.746	.16667	56	3136	7.483	23.664	.01786
7	49	2.646	8.367	.14286	57	3249	7.550	23.875	.01754
8	64	2.828	8.944	.12500	58	3364	7.616	24.083	.01724
9	81	3.000	9.487	.11111	59	3481	7.681	24.290	.01695
10	100	3.162	10.000	.10000	60	3600	7.746	24.495	.01667
11	121	3.317	10.488	.09091	61	3721	7.810	24.698	.01639
12	144	3.464	10.954	.08333	62	3844	7.874	24.900	.01613
13	169	3.606	11.402	.07692	63	3969	7.937	25.100	.01587
14	196	3.742	11.832	.07143	64	4096	8.000	25.298	.01562
15	225	3.873	12.247	.06667	65	4225	8.062	25.495	.01538
16	256	4.000	12.649	.06250	66	4356	8.124	25.690	.01515
17	289	4.123	13.038	.05882	67	4489	8.185	25.884	.01493
18	324	4.243	13.416	.05556	68	4624	8.246	26.077	.01471
19	361	4.359	13.784	.05263	69	4761	8.307	26.268	.01449
20	400	4.472	14.142	.05000	70	4900	8.367	26.458	.01429
21	441	4.583	14.491	.04762	71	5041	8.426	26.646	.01408
22	484	4.690	14.832	.04545	72	5184	8.485	26.833	.01389
23	529	4.796	15.166	.04348	73	5329	8.544	27.019	.01370
24	576	4.899	15.492	.04167	74	5476	8.602	27.203	.01351
25	625	5.000	15.811	.04000	75	5625	8.660	27.386	.01333
26	676	5.099	16.125	.03846	76	5776	8.718	27.568	.01316
27	729	5.196	16.432	.03704	77	5929	8.775	27.749	.01299
28	784	5.292	16.733	.03571	78	6084	8.832	27.928	.01282
29	841	5.385	17.029	.03448	79	6241	8.888	28.107	.01266
30	900	5.477	17.321	.03333	80	6400	8.944	28.284	.01250
31	961	5.568	17.607	.03226	81	6561	9.000	28.460	.01235
32	1024	5.657	17.889	.03125	82	6724	9.055	28.636	.01220
33	1089	5.745	18.166	.03030	83	6889	9.110	28.810	.01205
34	1156	5.831	18.439	.02941	84	7056	9.165	28.983	.01190
35	1225	5.916	18.708	.02857	85	7225	9.220	29.155	.01176
36	1296	6.000	18.974	.02778	86	7396	9.274	29.326	.01163
37	1369	6.083	19.235	.02703	87	7569	9.327	29.496	.01149
38	1444	6.164	19.494	.02632	88	7744	9.381	29.665	.01136
39	1521	6.245	19.748	.02564	89	7921	9.434	29.833	.01124
40	1600	6.325	20.000	.02500	90	8100	9.487	30.000	.01111
41	1681	6.403	20.248	.02439	91	8281	9.539	30.166	.01099
42	1764	6.481	20.494	.02381	92	8464	9.592	30.332	.01087
43	1849	6.557	20.736	.02326	93	8649	9.644	30.496	.01075
44	1936	6.633	20.976	.02273	94	8836	9.695	30.659	.01064
45	2025	6.708	21.213	.02222	95	9025	9.747	30.822	.01053
46	2116	6.782	21.448	.02174	96	9216	9.798	30.984	.01042
47	2209	6.856	21.679	.02128	97	9409	9.849	31.145	.01031
48	2304	6.928	21.909	.02083	98	9604	9.899	31.305	.01020
49	2401	7.000	22.136	.02041	99	9801	9.950	31.464	.01010
50	2500	7.071	22.361	.02000	100	10000	10.000	31.623	.01000

From *Statistical Inference* by Helen M. Walker and Joseph Lev. Copyright, 1953 by Holt, Rinehart and Winston, Inc. Reprinted by permission of Holt, Rinehart and Winston, Inc.

TABLE 10   Values of $n!$ and log $n!$

For an explanation and application see Chapter 2, Problem 5.
The values of $n!$ are given to five significant figures, and for $n \geq 9$ these values must be multiplied by a power of ten. This power is the raised number to the right of the five significant figures. For example, $15! \approx 13,077 \times 10^8$.

n	$n!$	log $n!$	n	$n!$	log $n!$	n	$n!$	log $n!$
1	1	.00000	26	$40,329^{22}$	26.60562	51	$15,511^{62}$	66.19065
2	2	.30103	27	$10,889^{24}$	28.03698	52	$80,658^{63}$	67.90665
3	6	.77815	28	$30,489^{25}$	29.48414	53	$42,749^{65}$	69.63092
4	24	1.38021	29	$88,418^{26}$	30.94654	54	$23,084^{67}$	71.36332
5	120	2.07918	30	$26,525^{28}$	32.42366	55	$12,696^{69}$	73.10368
6	720	2.85733	31	$82,228^{29}$	33.91502	56	$71,100^{70}$	74.85187
7	5,040	3.70243	32	$26,313^{31}$	35.42017	57	$40,527^{72}$	76.60774
8	40,320	4.60552	33	$86,833^{32}$	36.93869	58	$23,506^{74}$	78.37117
9	$36,288^{1}$	5.55976	34	$29,523^{34}$	38.47016	59	$13,868^{76}$	80.14202
10	$36,288^{2}$	6.55976	35	$10,333^{36}$	40.01423	60	$83,210^{77}$	81.92017
11	$39,917^{3}$	7.60116	36	$37,199^{37}$	41.57054	61	$50,758^{79}$	83.70550
12	$47,900^{4}$	8.68034	37	$13,764^{39}$	43.13874	62	$31,470^{81}$	85.49790
13	$62,270^{5}$	9.79428	38	$52,302^{40}$	44.71852	63	$19,826^{83}$	87.29724
14	$87,178^{6}$	10.94041	39	$20,398^{42}$	46.30959	64	$12,689^{85}$	89.10342
15	$13,077^{8}$	12.11650	40	$81,592^{43}$	47.91165	65	$82,477^{86}$	90.91633
16	$20,923^{9}$	13.32062	41	$33,453^{45}$	49.52443	66	$54,434^{88}$	92.73587
17	$35,569^{10}$	14.55107	42	$14,050^{47}$	51.14768	67	$36,471^{90}$	94.56195
18	$64,024^{11}$	15.80634	43	$60,415^{48}$	52.78115	68	$24,800^{92}$	96.39446
19	$12,165^{13}$	17.08509	44	$26,583^{50}$	54.42460	69	$17,112^{94}$	98.23331
20	$24,329^{14}$	18.38612	45	$11,962^{52}$	56.07781	70	$11,979^{96}$	100.07841
21	$51,091^{15}$	19.70834	46	$55,026^{53}$	57.74057	71	$85,048^{97}$	101.92966
22	$11,240^{17}$	21.05077	47	$25,862^{55}$	59.41267	72	$61,234^{99}$	103.78700
23	$25,852^{18}$	22.41249	48	$12,414^{57}$	61.09391	73	$44,701^{101}$	105.65032
24	$62,045^{19}$	23.79271	49	$60,828^{58}$	62.78410	74	$33,079^{103}$	107.51955
25	$15,511^{21}$	25.19065	50	$30,414^{60}$	64.48307	75	$24,809^{105}$	109.39461

Abridged from Mosteller, Rourke and Thomas: *Probability with Statistical Applications*, second edition, 1970, Addison-Wesley, Reading, Massachusetts.

TABLE 11   Percentile Values of R for $d.f.$ Degrees of Freedom when
$\rho = 0$

For an application, see Chapter 5, Problem 4a.

$d.f.$	$r_{.95}$	$r_{.975}$	$r_{.99}$	$r_{.995}$	$r_{.9995}$	$d.f.$	$r_{.95}$	$r_{.975}$	$r_{.99}$	$r_{.995}$	$r_{.9995}$
1	.988	.997	.9995	.9999	1.000	30	.296	.349	.409	.449	.554
2	.900	.950	.980	.990	.999	35	.275	.325	.381	.418	.519
3	.805	.878	.934	.959	.991	40	.257	.304	.358	.393	.490
4	.729	.811	.882	.917	.974	45	.243	.288	.338	.372	.465
5	.669	.754	.833	.874	.951	50	.231	.273	.322	.354	.443
6	.622	.707	.789	.834	.925	55	.220	.261	.307	.338	.424
7	.582	.666	.750	.798	.898	60	.211	.250	.295	.325	.408
8	.550	.632	.716	.765	.872	65	.203	.240	.284	.312	.393
9	.521	.602	.685	.735	.847	70	.195	.232	.274	.302	.380
10	.497	.576	.658	.708	.823	75	.189	.224	.264	.292	.368
11	.476	.553	.634	.684	.801	80	.183	.217	.256	.283	.357
12	.458	.532	.612	.661	.780	85	.178	.211	.249	.275	.347
13	.441	.514	.592	.641	.760	90	.173	.205	.242	.267	.338
14	.426	.497	.574	.623	.742	95	.168	.200	.236	.260	329
15	.412	.482	.558	.606	.725	100	.164	.195	.230	.254	.321
16	.400	.468	.542	.590	.708	125	.147	.174	.206	.228	.288
17	.389	.456	.528	.575	.693	150	.134	.159	.189	.208	.264
18	.378	.444	.516	.561	.679	175	.124	.148	.174	.194	.248
19	.369	.433	.503	.549	.665	200	.116	.138	.164	.181	.235
20	.360	.423	.492	.537	.652	300	.095	.113	.134	.148	.188
22	.344	.404	.472	.515	.629	500	.074	.088	.104	.115	.148
24	.330	.388	.453	.496	.607	1000	.052	.062	.073	.081	.104
25	.323	.381	.445	.487	.597	2000	.037	.044	.016	.058	.074
	$-r_{.05}$	$-r_{.025}$	$-r_{.01}$	$-r_{.005}$	$-r_{.0005}$		$-r_{.05}$	$-r_{.025}$	$-r_{.01}$	$-r_{.005}$	$-r_{.0005}$

Taken from Fisher and Yates: *Statistical Tables for Biological, Agricultural and Medical Research*, published by Longman Group, Ltd., London (previously published by Oliver & Boyd, Edinburgh), and by permission of the authors and publishers.
$d.f.$ = degrees of freedom.

TABLE 12   Values for Transforming R into $z = \dfrac{1}{2} \ln \left(\dfrac{1 + R}{1 - R}\right)$

For an explanation and application see Chapter 4, Problem 4b.

z	.00	.01	.02	.03	.04	.05	.06	.07	.08	.09
.0	.0000	.0100	.0200	.0300	.0400	.0500	.0599	.0699	.0798	.0898
.1	.0997	.1096	.1194	.1293	.1391	.1489	.1587	.1684	.1781	.1878
.2	.1974	.2070	.2165	.2260	.2355	.2449	.2543	.2636	.2729	.2821
.3	.2913	.3004	.3095	.3185	.3275	.3364	.3452	.3540	.3627	.3714
.4	.3800	.3885	.3969	.4053	.4136	.4219	.4301	.4382	.4462	.4542
.5	.4621	.4700	.4777	.4854	.4930	.5005	.5080	.5154	.5227	.5299
.6	.5370	.5441	.5511	.5581	.5649	.5717	.5784	.5850	.5915	.5980
.7	.6044	.6107	.6169	.6231	.6291	.6352	.6411	.6469	.6527	.6584
.8	.6640	.6696	.6751	.6805	.6858	.6911	.6963	.7014	.7064	.7114
.9	.7163	.7211	.7259	.7306	.7352	.7398	.7443	.7487	.7531	.7574
1.0	.7616	.7658	.7699	.7739	.7779	.7818	.7857	.7895	.7932	.7969
1.1	.8005	.8041	.8076	.8110	.8144	.8178	.8210	.8243	.8275	.8306
1.2	.8337	.8367	.8397	.8426	.8455	.8483	.8511	.8538	.8565	.8591
1.3	.8617	.8643	.8668	.8693	.8717	.8741	.8764	.8787	.8810	.8832
1.4	.8854	.8875	.8896	.8917	.8937	.8957	.8977	.8996	.9015	.9033
1.5	.9052	.9069	.9087	.9104	.9121	.9138	.9154	.9170	.9186	.9202
1.6	.9217	.9232	.9246	.9261	.9275	.9289	.9302	.9316	.9329	.9342
1.7	.9354	.9367	.9379	.9391	.9402	.9414	.9425	.9436	.9447	.9458
1.8	.9468	.9478	.9498	.9488	.9508	.9518	.9527	.9536	.9545	.9554
1.9	.9562	.9571	.9579	.9587	.9595	.9603	.9611	.9619	.9626	.9633
2.0	.9640	.9647	.9654	.9661	.9668	.9674	.9680	.9687	.9693	.9699
2.1	.9705	.9710	.9716	.9722	.9727	.9732	.9738	.9743	.9748	.9753
2.2	.9757	.9762	.9767	.9771	.9776	.9780	.9785	.9789	.9793	.9797
2.3	.9801	.9805	.9809	.9812	.9816	.9820	.9823	.9827	.9830	.9834
2.4	.9837	.9840	.9843	.9846	.9849	.9852	.9855	.9858	.9861	.9863
2.5	.9866	.9869	.9871	.9874	.9876	.9879	.9881	.9884	.9886	.9888
2.6	.9890	.9892	.9895	.9897	.9899	.9901	.9903	.9905	.9906	.9908
2.7	.9910	.9912	.9914	.9915	.9917	.9919	.9920	.9922	.9923	.9925
2.8	.9926	.9928	.9929	.9931	.9932	.9933	.9935	.9936	.9937	.9938
2.9	.9940	.9941	.9942	.9943	.9944	.9945	.9946	.9947	.9949	.9950
3.0	.9951									
4.0	.9993									
5.0	.9999									

Taken from Fisher and Yates: *Statistical Tables for Biological, Agricultural and Medical Research*, published by Longman Group, Ltd., London (previously published by Oliver & Boyd, Edinburgh) and by permission of the authors and publishers. The figures in the body of the table are values of R corresponding to z-values read from the scales on the left and top of the table.

Subject Index

259